HYBRID AND BATTERY ENERGY STORAGE SYSTEMS
REVIEW AND RECOMMENDATIONS FOR PACIFIC ISLAND PROJECTS

AUGUST 2022

IAN DEVELOPMENT BANK

© 2022 Asian Development Bank
6 ADB Avenue, Mandaluyong City, 1550 Metro Manila, Philippines
Tel +63 2 8632 4444; Fax +63 2 8636 2444
www.adb.org

Some rights reserved. Published in 2022.

ISBN 978-92-9269-661-0 (print); 978-92-9269-662-7 (electronic); 978-92-9269-663-4 (ebook)
Publication Stock No. TCS220320-2
DOI: http://dx.doi.org/10.22617/TCS220320-2

Note:
In this publication, "$" refers to United States dollars.

Cover design by Edith Creus. On the cover: Tonga, Tongatapu, Popua Power Station Maama Mai Solar PV and BESS (Top); and Cook Islands, Aitutaki, Power Station Solar PV (Bottom) (Photos by TPL and Entura).

Contents

Tables and Figures

Foreword

A transition from imported diesel-based power generation toward locally available renewable energy generation has been a national priority of Pacific Small Island Developing States (Pacific SIDS). Through this transition they aim to address multiple climate change challenges and ensure energy security and greater energy access.

The Asian Development Bank (ADB) supports its 14 Pacific SIDS clients in their journeys to a renewable energy future by providing financing and enabling the use of advanced technologies, such as battery energy storage systems (BESS), in ADB-funded projects. Notable examples include projects in the Kingdom of Tonga (Tonga) and the Cook Islands. These countries have become early technology adopters: their relatively small size and ambitious targets help in the quick transition to high renewable energy penetration.

This report reviews several ADB-funded projects as case studies to assess and better understand the success factors and opportunities to improve the implementation of renewable energy-based hybrid electricity systems with battery energy storage systems. The assessment focuses on the technological, procurement and contract management and the way decision-making processes are applied in these projects to achieve success. The lessons learned and proposed practical recommendations of the assessment could guide policymakers, power utilities and the private sector in initiating and implementing similar projects in other countries of the Pacific region.

ADB's assistance to the energy sector of the Pacific is helping to build resilience to climate change and external shocks, improve sustainable service delivery and expand access to renewable energy, and support private sector growth. ADB's Pacific energy sector operations strengthen energy security, ensure a cleaner environment by reducing greenhouse gas emissions and pollution from diesel-based generation and oil spills, and pave the way for a more prosperous Pacific community.

ADB is committed to supporting Pacific countries in building resilient energy infrastructure and making successful transitions to renewable energy. The bank will closely work with governments, communities, and development partners to help achieve national energy sector priorities and goals.

Leah Gutierrez
Director General
Pacific Department
Asian Development Bank

Acknowledgments

This report was jointly prepared by Woo Yul Lee, Principal Energy Specialist, Energy Division, Pacific Department (PAEN) and Chris Blanksby, Specialist Renewable Energy Engineer, Entura, Australia.

We wish to acknowledge the guidance provided by Mukhtor Khamudkhanov, Director, PAEN and contributions from Pacific Department colleagues: Eun Young So, Energy Specialist, PAEN; Ranishka Yasanga Wimalasena, Energy Specialist, Pacific Subregional Office in Suva, Fiji; and Maria Regina P. Arquiza, Operations Assistant, PAEN.

We also acknowledge the valuable contributions to the content of this report by the many project stakeholders in the Cook Islands and Tonga who, with the assistance of Simon Wilson and Stephen Lyon, took the time to share their views and insights, suggestions, photos, and expertise with the authors.

We would also like to thank Dae Kyeong Kim, a peer reviewer of this report and a consultant in ADB's Sustainable Development and Climate Change Department.

Finally, we would like to thank the following cofinanciers for the projects used as case studies for this report.

(i)   European Commission, Global Environment Facility, and Green Climate Fund (Renewable Energy Sector Project in the Cook Islands);

(ii)   European Commission, Global Environment Facility, Second Danish Cooperation Fund for Renewable Energy and Energy Efficiency in Rural Areas and Government of Australia (Outer Islands Renewable Energy Project in Tonga);

(iii) Green Climate Fund and Government of Australia (Renewable Energy Project in Tonga).

Cecilia C. Caparas, Associate Knowledge Management Officer, Pacific Department, oversaw the publication process. The report was edited by Jess Macasaet and proofread by Cherry Lynn Zafaralla. Edith Creus executed the layout and design.

Abbreviations

ADB	Asian Development Bank
AUX	auxiliary
BESS	battery energy storage system
BMS	battery management system
BTM	behind-the-meter
CIRESP	Cook Islands Renewable Energy Sector Project
COVID-19	coronavirus disease
DFAT	Department of Foreign Affairs and Trade, Australia
DLP	defects liability period
ECI	early contractor involvement
EMS	energy management system
EPC	engineering, procurement, and construction
GCF	Green Climate Fund
GEF	Global Environment Facility
HMI	human–machine interface
IPP	independent power producer
kW	kilowatt
kWh	kilowatt-hour
MW	megawatt

MWh	megawatt-hour
NDC	nationally determined contribution
O&M	operation and maintenance
OIREP	(Tonga) Outer Island Renewable Energy Project
PCU	power conversion unit
PPA	power purchase agreement
SIDS	small island developing states
SoC	state of charge
TA	technical assistance
TREP	Tonga Renewable Energy Project

Executive Summary

The 14 Pacific small island developing states (Pacific SIDS) clients of the Asian Development Bank (ADB) are among the smallest and most remote countries in the world. Many Pacific SIDS have set ambitious targets for renewable energy uptake, focused initially on the electricity sector. For instance, the Cook Islands (COO) planned for 100% of islands to be powered by renewable electricity by 2020, while the Kingdom of Tonga (TON or Tonga) targeted 50% electricity from renewable energy by 2020 and 70% by 2030.

These objectives are driven by circumstances and challenges unique to these countries.

(i)   The Pacific SIDS have some of the highest electricity prices globally, averaging $0.38 per kilowatt-hour (kWh).

(ii) The population in the outer islands has low electrification rates, with some relying on portable generators or solar home systems that provide an unreliable or expensive solution.

(iii) The Pacific SIDS are also significantly more vulnerable to a range of external factors that impact the security and cost of energy supply. These include constrained supply chains for fuel; undiversified economies that increase exposure to external shocks; and geological exposure to sea-level rise, storm surge, and other extreme events exacerbated by climate change.

(iv) Coupled with these factors, many Pacific SIDS lack infrastructure, capacity, and services to provide resilience against such impacts.

Many Pacific SIDS see a conversion of imported fossil fuel-based electricity generation to locally available renewable energy generation as an opportunity to reduce vulnerability, increase sustainability and energy security, and improve electricity access and affordability. With the assistance of development partners like ADB, renewable energy projects initiated since 2014 have strived to achieve these outcomes. In the Cook Islands and Tonga, the following projects funded by ADB (and other development partners) were established:

(i)   Cook Islands: Renewable Energy Sector Project;[1]

(ii) Tonga: (i) Outer Islands Renewable Energy Project;[2] and (ii) Renewable Energy Project.[3]

---

[1]   Cofinanced by the European Commission (EC), Global Environment Facility (GEF), and Green Climate Fund (GCF).
[2]   Cofinanced by the Government of Australia, EC, Government of Denmark, and GEF.
[3]   Cofinanced by EC, GCF, and the Government of Australia.

These projects cover a total of 26 subprojects over 19 islands, with six independent lead contractors representing a broad range of scale and conditions for transition, including from:

(i)   Solar and battery microgrids, providing electrification to islands with populations under 100; to

(ii)   Integration of multi-megawatt scale distributed solar, wind, and storage into an existing diesel generation based network serving the population of close to 90,000.

These projects are not without challenges. Underlying many of the challenges is that these countries have become early technology adopters: their relatively small size and ambitious targets mean that transition to renewables can occur quickly. In addition, the necessary technology for managing variable renewable energy is maturing and changing rapidly.

This study reviews these projects as case studies to assess and better understand the success factors and opportunities to improve the implementation of battery energy storage systems (BESS) and renewable energy-based hybrid electricity systems. The assessment focus on the technological, procurement and contract management, and decision-making process applied in those projects.

The methodology used in conducting this review is adapted from a standard project post implementation review and applied across a portfolio of projects. The authors led the review, relying on the core project delivery team and key stakeholder representatives as the subject matter experts. Interviews were conducted with these stakeholders to understand and evaluate project performance and key issues or concerns.

This methodology resulted in identification of four principal areas of challenge for the projects:

- Technical challenges:
  - Maturing BESS industry
  - Improving alignment between technical considerations and decision-making
- Standards, safety and environmental challenges
- Procurement challenges:
  - Constrainted procurement options
  - Aligning interrelated contracts
  - Managing contractor incentives.
- COVID-19 challenges:
  - Insurance costs and availability
  - Remote commissioning.

The authors explore each area to determine the factors affecting project performance and potential opportunities to mitigate these. They draw on a review of the role of hybrid electricity systems for isolated networks, a literature review on BESS and hybrid technology, consideration of procurement approaches, and detailed case study descriptions.

This assessment is used to develop recommendations that focus on opportunities for improvement in project performance or for mitigating risk factors. These recommendations are described below.

Project design

(i) There are significant knowledge gaps for stakeholders about the associated technical issues, particularly with medium to high renewable energy hybrid systems, BESS, technology selection, and control systems requirements. Technical assistance consultants cannot always understand or address the drivers and needs of stakeholders. These factors contribute to reduced accuracy of risk assessments and sub- optimal decision-making, which can be addressed through a consolidated program to build and maintain local energy literacy, supported by tools and information designed to communicate key concepts clearly. Some examples are presented in this report.

(ii) Information to support decision-making is insufficient without a structured model for informed decision-making such as the responsible-accountable, consult, inform (RACI) model. For hybrid energy projects in isolated grids, stakeholder engagement during initial project selection and definition may benefit from utilizing the following model:

   (a) **Sponsor:** Allocate based on country structure (likely utility, government energy department)
   (b) **Government (energy ministry):** Responsible and accountable - strategic decision-maker
   (c) **Government (financial ministry):** Consult
   (d) **Utility:** Consult
   (e) **Regulator:** Inform
   (f) **Customers:** Consult and inform
   (g) **Landowners:** Consult and inform
   (h) **Regional developers, contractors, and investors:** Inform
   (i) **Project management unit (PMU), technical assistance:** Consult

(iii) Where possible, consistent delivery teams, including project management, administration, and technical assistance to support a unified decision-making team are also recommended.

Technology

(i) Technology has matured substantially since the case study projects commenced. It is now apparent that BESS can offer a full suite of grid support functions allowing stable operation of small, medium and large isolated networks with high renewable contribution, even without synchronous (diesel) generation online. However, there is still significant progress to be made, and particular gaps remain in these areas:

   (a) product standardization,
   (b) end-of-life treatment, including replacement,
   (c) clarity on emergency services response requirements, and
   (d) consistency in definitions of control capabilities and in diesel-off operational capability.

(ii)   For future projects, it is critical to understand these gaps. Noting that BESS products are not yet highly standardized, specifiers must give detailed consideration to the required project-specific functionality and operating environment and specify or select applicable standards or requirements accordingly.

(iii) It is also recommended to monitor ongoing technology advancement, including standards, and apply continuous improvement to technical specifications and concept development. However, presently this may be demanding for Pacific SIDS and small utilities and they should be supported through technical assistance or funding partners in the short to medium term.

Procurement

(i)   Procurement processes were identified as a challenge for many stakeholders. There was a desire to consider risk and opportunity through merit-based evaluation (particularly important for small, customized projects in remote locations, in a market with limited competition), or to manage complex, innovative projects. This approach is now facilitated in ADB's 2017 Procurement Policy that is applicable to all new projects using ADB financing. The following is recommended to allow better adaptation for hybrid projects:

   (a)   Undertake a comprehensive strategic procurement planning (SPP) exercise during the feasibility stage of a project in parallel with its technical development to identify an optimal procurement strategy that will deliver value-for-money outcomes.

   (b)   Include merit-point assessment criteria in the evaluation of all complex tenders, as standard, unless the SPP exercise determines it to not be the most suitable approach.

   (c)   Consider all available contracting modalities (e.g. Early Contractor Involvement – ECI) when developing the procurement strategy, ensuring that the modality chosen is best suited to the project and will facilitate effective competition.

   It is recommended that the project delivery team engage early with ADB to utilize the flexibilities in the 2017 Procurement Policy.

(ii)   Additionally, given the nonstandard nature of projects to date and relatively high-risk exposure of the employer through to completion and commissioning, it is recommended to consider slight changes to performance securities and payment milestones. In particular, payment milestones should consider higher completion and commissioning payments. Performance securities should be maintained at a higher level through the first 2 years of operation, while battery degradation is verified. However, protections under the contract will also rely on strong project management processes that enable enforcement of the relevant protection measures.

(iii) For projects requiring alignment, particularly where BESS or other utility-owned technology was deployed to support connection of independent power producers (IPPs), it was considered most advantageous to plan for completing BESS at 3-9 months ahead of the IPPs. This was found to provide reasonable mitigation against the more significant risk of delaying the start of IPPs' commercial operations since in many cases, the BESS can still provide some project benefits prior to IPP connection.

(iv)   Finally, in light of potential ongoing disruptions to travel for Pacific SIDS, it is recommended that all contracts contain a provision for remote commissioning and servicing from regional locations.

Insurance

(i)   Insurance options for Pacific projects are currently very limited. Considering their portfolio of investment and interest, an ADB or other funding agency backed insurance scheme may be a viable alternative. This would not only offer potential savings but reduce administrative time and cost in sourcing and negotiating insurance on a project by project basis. However, issues such as the impact on market competition for insurance services or fit with the long-term operations period of projects would require careful consideration.

Cook Islands, Aitutaki, Power Station Solar PV
(Photo by Entura).

Introduction

The 14 Pacific small island developing states (Pacific SIDS)[1] clients of the Asian Development Bank (ADB) are among the smallest and most remote countries in the world, with a combined population of approximately 10 million people. Many Pacific SIDS have set ambitious targets for renewable energy uptake, focussed initially on the electricity sector. For instance, the Cook Islands planned for 100% of islands to be powered by renewable electricity by 2020, while the Kingdom of Tonga (Tonga) targeted 50% electricity from renewable energy by 2020 and 70% by 2030. These objectives are driven by circumstances and challenges unique to these countries.

Pacific SIDS have some of the highest electricity prices globally, averaging $0.38 ranging from $0.18/kWh to $0.58/kWh (Utilities Regulatory Authority, 2019). The population in the outer islands has low electrification rates, relying on portable generators or solar home systems that provide an unreliable or expensive solution. These countries are also significantly more vulnerable than other countries that impact the security and cost of energy supply (United Nations Development Programme 2021). Such factors include constrained supply chains: a shortage of diesel fuel for electricity generation has previously occurred due to shipping delays. Another factor is undiversified economies that increase exposure to external shocks. This exposure was strongly exemplified by the virtual cessation of tourism in the Cook Islands during the coronavirus disease (COVID-19) pandemic. Tourism had previously accounted for nearly 70% of the country's gross domestic product (GDP) (Syme-Buchanan 2019). A further factor is that many Pacific SIDS are also geologically low-lying and exposed to sea level rise, storm surge, and other extreme events exacerbated by climate change.

---

[1] Cook Islands, the Federated States of Micronesia, Fiji, Kiribati, Nauru, Niue, Palau, Papua New Guinea, the Marshall Islands, Samoa, Solomon Islands, Tonga, Tuvalu, and Vanuatu.

Coupled with these factors, many Pacific SIDS lack infrastructure, capacity, and services to provide resilience against such impacts.

Many Pacific SIDS see a conversion of imported fossil fuel-based electricity generation to locally available renewable energy generation as an opportunity to reduce vulnerability by addressing the multiple challenges of climate change, energy security, and energy access; see, for example, from the Cook Islands, Ministry of Finance and Economic Management (2017). With the assistance of development partners like ADB, progress toward these targets since 2014 has been encouraging, but not without challenges. Underlying many of the challenges is that these countries have become early technology adopters: their relatively small size and ambitious targets mean that transition to high renewables can occur quickly. This occurs in an environment where the necessary technology for managing variable renewable energy matures and changes rapidly.

With limited hydropower opportunities in many Pacific SIDS, the typical path for transformation requires integrating distributed intermittent solar and wind generators, energy storage, and existing associated infrastructure. In the Cook Islands and Tonga, the following projects funded by ADB were established to implement this approach:

  (i)   Cook Islands: Renewable Energy Sector Project;[2]

  (ii)   Tonga: (i) Outer Islands Renewable Energy Project;[3] and (ii) Renewable Energy Project.[4]

These projects cover a total of 26 subprojects over 19 islands, with six independent lead contractors representing a broad range of scale and conditions for transition, including from:

  (i)   Solar and battery microgrids, providing electrification to islands with populations under 100; to

  (ii)   Integration of multi-megawatt (MW) scale distributed solar, wind, and storage into an existing diesel generation based network serving the population of close to 90,000.

These projects are used here as case studies to assess and better understand the success factors and opportunities to improve the implementation of battery energy storage systems (BESS) and renewable energy-based hybrid electricity systems. The assessment is focused on the technological, procurement and contract management, and decision-making process applied in those projects, and recommendations are made for Pacific SIDS and other regions facing similar challenges.

1.1   Report Structure

This report initially provides a general overview of BESS and hybrid renewable electricity systems for small electricity grids in section 2. This is intended to provide context for the case studies and essential background on the challenges of the transition to high renewable energy contribution power systems. Section 3 then describes the case studies that are used as the basis for this review, using each project's stated objectives. This is followed by a description of the scope, method of review, and inputs used for the study in section 4. Finally, the core themes identified through the investigation are presented in section 5 along with an analysis of the opportunities and risks they present for future projects. The key recommendations are then summarized in section 6.

---

[2]   Cofinanced by the European Commission (EC), Global Environment Facility (GEF), and Green Climate Fund (GCF).
[3]   Cofinanced by the Government of Australia, EC, Government of Denmark, and GEF.
[4]   Cofinanced by EC, GCF, and the Government of Australia.

1.2 Study Limitations

This assessment focuses only on the "implementation" aspects (listed in section 4) of BESS and hybrid renewable energy projects. However, important lessons can also be derived from related operation experiences arising from these projects that are not covered here. This includes methods for estimating the economic benefit of BESS, renewable energy installation in Pacific SIDS, performance assessment of BESS and hybrid projects, operation and maintenance (O&M) requirements of BESS and hybrid projects, financing models and opportunities for increasing renewable energy in Pacific SIDS, and asset management of BESS. This assessment is intended to be followed by an evaluation of performance assessment and O&M requirements once the systems have been in operation for a longer period.

This study is also limited to the lessons learned through the specified case studies. In the authors' experience, these projects represent similar projects undertaken elsewhere, so the findings are likely to be broadly applicable to other SIDS facing similar challenges. Nevertheless, the scope and context of each project are different, and recommendations presented here may not be applicable in all circumstances.

2

Tonga, Tongatapu, Power Station Grid Stability BE
(Photo by TPL).

Role of Hybrid Electricity Systems for Isolated Networks

Hybrid electricity systems describe the integration of multiple technologies, typically emerging and renewable energy technology, to deliver electricity to customers safely, reliably, and efficiently. Hybrid electricity systems are differentiated from conventional ones that typically utilize one technology for centralized generation for distribution to customers via an electricity network or grid. A comparison of the elements of a typical conventional and hybrid electricity system for an isolated network[5] is shown in Figure 2.1.

In Tongatapu, the main island of Tonga, for example, the previous conventional electricity system consisted of a centralized diesel generator power station, with radial medium-voltage (MV) electrical feeders and low-voltage (LV) grids to distribute this power to customers. This is transitioning to a hybrid electricity system, which retains the conventional elements but adds intermittent renewable generation embedded within the MV network (independent medium-scale wind or solar projects) and LV network (customer rooftop solar), has bidirectional power flows, adds new network communications and control capabilities, and adds storage to manage variability in renewable generation.

---

[5] The Cook Islands and Tonga operate "isolated" networks, and this is typical of all Pacific SIDS. An isolated network is one where there is a primary generation location (power station, potentially with multiple generators) that must be operating to maintain power to all connected loads in that network. There are no interconnections to other networks with their own generation sources that can maintain power to load in the network if the primary generation location is unavailable. Isolated networks can include multiple distributed generation and storage sources such as solar photovoltaic (PV), BESS, or wind.

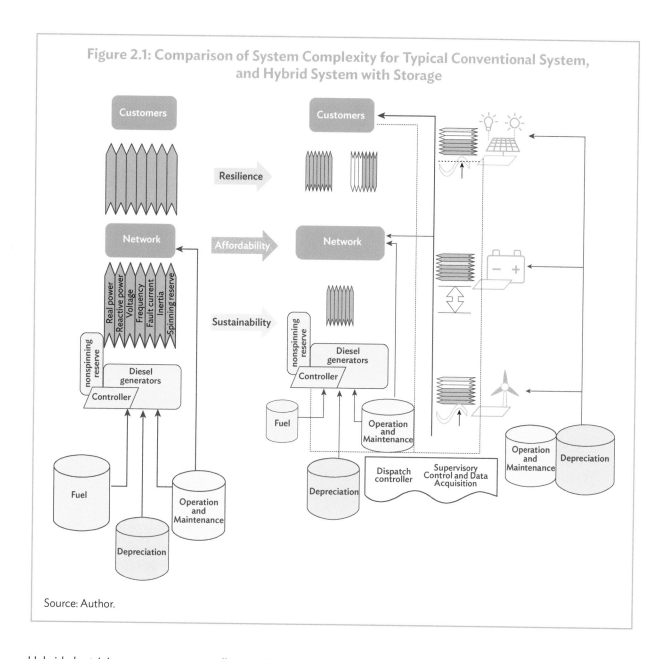

Figure 2.1: Comparison of System Complexity for Typical Conventional System, and Hybrid System with Storage

Source: Author.

Hybrid electricity systems are typically introduced in response to one or more of four drivers:

(i)   **Electrification**—for Outer Islands with no existing power supply, a hybrid solution is often the lowest cost solution to electrify villages. The renewable energy offered by hybrid systems is also typically aligned with priorities for funding agencies that may be relied upon to support the electrification of outer islands.

(ii) **Sustainability**—primarily reduced greenhouse gas emissions.

(iii) **Affordability**—potential to save fuel costs and O&M costs over the project life of a hybrid system. This driver may include the perceived affordability benefit, but not all projects may offer an increase in affordability.

(iv) **Resilience**—less reliance on fuel imports and price volatility. This includes reducing vulnerability to supply chain interruptions during severe weather or geopolitical events.

The transition from conventional to hybrid systems can be measured by the percentage contribution of renewable energy to the annual customer load. Low renewable contributions can accommodate renewable generation with minimal modification to the conventional infrastructure. Still, control, storage, and other significant changes to infrastructure and operations are required as renewable contribution increases. As Figure 2.1 shows, the complexity of the full hybrid system is considerably higher than the conventional system.

Some of the important elements shown in Figure 2.1 are the following:

(i)   Electricity systems require more than just real power to operate. Other system functions traditionally delivered natively by diesel generators are

    (a)   reactive power,

    (b)   voltage control,

    (c)   frequency control,

    (d)   fault current,

    (e)   inertia, and

    (f)   spinning reserve.[6]

(ii)   Not all of the above functions can be provided by renewable generators. For instance, Figure 2.1 shows that solar PV can only provide real and reactive power, voltage control, and limited spinning reserve (arrows in green are the services provided). BESS has recently developed the capability to provide all these functions. However, they may not represent the least cost, particularly with respect to the provision of fault current. Thus, hybrid combinations, integration, and coordination of functions from different technologies are required (for example, provision of frequency control from a BESS if diesel generators are not active, even if solar supplies all the real power required by customers).

(iii)   Solar, wind, and storage all have additional considerations in their operation:

    (a)   Solar and wind are variable energy sources and may exceed the requirement, with excess energy-requiring storage or curtailment.

    (b)   BESS provides energy that is dispatchable on demand but is limited by their storage capacity and must be managed within the state of charge limits.

(iii)   Power flows are no longer unidirectional, from the power station via the network to customers. Instead, power may flow into the network from various distributed generators or may be supplied and consumed by customers.

By way of comparison with a conventional diesel-based system, a hybrid system introduces the following elements to the system operators (typically utilities) that provide new levels of complexity for them to learn and manage:

(i)   More individual system components to control and manage (see Figure 2.1).

(ii)   More types of system components and their individual characteristics (see Figure 2.1).

(iii)   System components using advanced technology, such as inverter-based energy systems that rely on very high-speed computerized control systems and power electronics, or lithium batteries that require cell level monitoring of thousands of data points for normal operations.

---

[6]   Nonspinning reserve is also an important feature of diesel generators, allowing additional generating units to be brought online within some specified time period. BESS and solar PV also can offer nonspinning reserve; however, the very low marginal cost of maintaining these assets online means that unlike diesel generators, their capacity is typically maintained online at all times (except during maintenance).

(iv)   System components that are still maturing as a technology.

(v)   Requirement for communication and coordination of functions between system components.

(vi)   Multiple providers of energy, introducing new commercial relationships.

Figure 2.2: The Energy Trilemma Challenge—Increasing Sustainability while Maintaining Reliability and Affordability

Renewable generation benefits sustainability. It also provides energy at lower cost but reduces system reliability.

Storage adds reliability but also adds cost. Balancing storage with renewable generation to maintain affordability and reliability is the key objective of the energy trilemma.

Source: Author.

Inevitably, the transition to hybrid systems with storage gives rise to the "energy trilemma" illustrated in Figure 2.2. In the context of the energy trilemma, renewable energy is cheaper than traditional diesel-based generation. However, increasing the amount of renewable energy changes the variability of supply and the mismatch between supply and demand, and introduces requirements for supplementary technologies to maintain reliability. The cost of these additional technologies increases the effective cost of renewable energy. A balance means finding a level of renewables that offers a risk-weighted least-cost solution and offers flexibility to expand in the future as opportunities arise (Nikolic et al. 2016).

The timescale over which variability and potential mismatch in supply and demand occur is important to the function of a BESS. The different scenarios are highlighted in Table 2.1.

Table 2.1: Battery Energy Storage System Applications at Various Timescales

Storage capacity	Cycle interval	Cause of Variability	Function	Descriptor
Short	Millisecond to minute	Fault, or fast change in resource such as cloud bank for solar PV or wind gust for a wind turbine	Fault current, frequency and voltage setting and/or support, spinning reserve	Grid support
Medium	Hour	Time difference between midday peak solar PV generation and evening peak load	Reliable peak demand	Energy arbitrage
Long	Day	Day–night solar change. Extended cloudy periods	High renewables	Load shifting

PV = photovoltaic.
Source: Author.

2.1 Role of Battery Energy Storage System in Hybrid Electricity Systems

Battery energy storage systems now fill a critical role in enabling hybrid energy systems at higher levels of renewable energy contribution because they offer two key capabilities in a moderately robust and cost-effective package:

(i)   Short- to long-term storage capacity (including load-shifting) and associated functions are often described as load-shifting BESS (see Figure 2.3).

(ii) All power system functions listed in section 2.1 are often described as a grid stability BESS.[7]

Because of this, BESS can form a stable and operational grid using renewable energy without diesel generation. BESS features in all of the hybrid systems implemented in the Cook Islands and Tonga. However, BESS have limitations:

(i)   Functionality can be limited depending on the state of charge (SoC) of the BESS. If full, the BESS can provide power to support the grid but cannot accept any excess energy, and therefore may allow the frequency to increase. If the BESS can accept excess energy at minimum SoC but cannot provide power to support the grid, and frequency may decrease. It is necessary to operate BESS leaving some reserve capacity below full (around 5%) and above empty (10%–20%).

(ii) While rapidly approaching maturity, BESS still lag established power system technologies (such as diesel, solar, wind, cables, transformers or switchgear), particularly in aspects such as standardization, operating procedures, or available operational performance data.

(iii) Despite substantial cost decline and filling a critical niche, BESS is still a significant cost contributor to power systems. Therefore, optimising sizing for the least cost is critical, limiting the viable threshold for the percentage contribution of renewable energy.

Battery energy storage systems can be employed in different roles in a hybrid electricity system (this is discussed in more detail in section 5.1.2, however, at a high level):

(i)   Behind-the-meter (BTM) BESS is small system, typically with up to 10 kW power capacity, located at a customer's premises (generally on the customer side of the revenue meter and owned by the customer), suitable for managing an individual customers needs. They are also required to provide basic network support functions.

(ii) Utility-scale BESS is larger system, typically with a power capacity of the same order of magnitude as the network or local feeder peak load. They are located at the central power station or at strategic locations in the network (which may be at a renewable generation site). Their primary role is in supporting the utility to manage energy and power flow in the network. Therefore, communication and integration with the utility control system are essential.

In the context of this investigation, utility scale BESS is more critical and representative of the case studies examined.

---

[7]   This was not the case historically; see section 5.1.1.

2.2 Impact of Scale of Hybrid System

Three scales of the electricity system are discussed in this study. These have different equipment characteristics and implementation strategies and align well with the scale of islands in many Pacific SIDS. Typical parameters for these are as follows.

2.2.1 Small-Scale Systems

(i)   Peak load of the system is 10–100 kW.

(ii)   Typically a single, centralized generation plant (may be greenfield site with no existing electrification prior to project implementation).

(iii) Low-voltage distribution network to populations of less than 1,000.

(iv)   Typical examples: Outer islands

    (a)   **Tonga:** Niuafo'ou, Niuatopatapu, 'Uiha, Nomuka, Ha'ano, Ha'afeva, Kotu, Tungua, O'ua, Mo'unga'one (and others not covered under the case studies).

    (b)   **Cook Islands:** Atiu, Mangaia, Mauke, Mitiaro (and others not covered under the case studies).

(iii) Example combination of technology:

    (a)   Solar PV: 70 kW

    (b)   BESS 40 kW / 600 kWh

    (c)   Diesel Generator 2 x 40 kW

(vi)   High levels of renewable contribution can be achieved more easily in small systems with a combination of off-the-shelf technologies, low maintenance needs, and less critical reliability requirements (Nikolic et al 2016). The cost of energy is typically high due to scale and remoteness but may be less than a diesel- only system. For the case study projects, the cost of energy for these cases was typically above 1.0, at least twice the cost of energy of medium- or large-scale systems.

Mauke Hybrid Power Station in the Cook Islands. An example of a small-scale project (photo by Entura).

(vi)   Due to the higher cost of energy and low ability to pay, projects are typically delivered as fully grant-funded or subsidized by larger grids. Cost recovery of O&M over the life of the project may be the only requirement to demonstrate financial sustainability. Execution is typically via an engineering, procurement, and construction (EPC) contract with ownership residing with government (through a utility or other entity), consolidating several small projects as the projects are too small for independent power producer (IPP) contractors. Small-scale and high mobilization costs mean most projects are delivered in a single stage as compared to larger systems where renewable energy and BESS may be installed in a number of stages.

2.2.2 Medium-Scale Systems:

(i)   Peak load of the system is 100-1,000 kW.

(ii)   Typically a centralized generation plant, often with multiple renewable energy sites embedded in the network.

(iii)   MV distribution network to populations of up to 10,000.

(iv)   Typical examples: Larger populated islands and secondary population centers.

    (a)   **Tonga:** Vava'u, 'Eua and Ha'apai.

    (b)   **Cook Islands:** Aitutaki.

(v)   Typical combination of technology

    (a)   Solar: 750 kW

    (b)   BESS: 500 kW / 500 kWh

    (c)   Diesel: 2 x 600 kW + 1 x 300 kW

(vi)   Medium-scale projects are suited for implementation in 2–4 stages of discrete infrastructure projects, progressively adding renewable generation with storage and control upgrades. Staging implementation can help to maintain financial viability as technology costs decrease over time. Staging also allows operators time and experience to adapt to changes and gain operational experience to ensure successive stages are optimally designed as the renewable energy contribution increases and complexity of operation increases.

(v)   Renewable generation is typically delivered together with necessary storage and control under a single EPC contract. The modest scale of these projects is perceived as limiting commercial attractiveness for other delivery models such as a power purchase agreement (PPA).

Figure 2.3: Example of Medium-Scale Project (Aitutaki, Cook Islands) (750 kW solar PV and 500 kW / 500 kWh BESS)

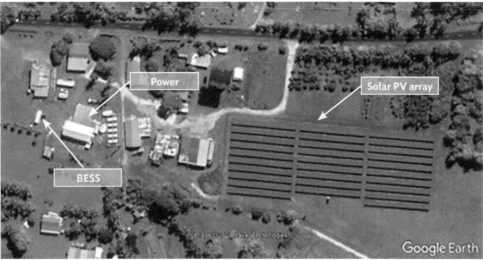

BESS = battery energy storage system, kW = kilowatt, kWh = kilowatt-hour, PV = photovoltaic.
Note: 750 kilowatt solar photovoltaic plant and 500 kilowatt-hour BESS.
Source: Google Earth.

2.2.3 Large-Scale Systems

(i)   Peak load of the system is 1–10 MW.

(ii)   Typically a centralized power station, multiple renewable generators and some storage embedded in the network, BTM generation and storage.

(iii) Medium-voltage distribution network to popluations of up to 100,000.

(iv)   Island suitability: Main island

    (a)   **Tonga:** Tongatapu

    (b)   **Cook Islands:** Rarotonga

(v)   Large scale projects are multifaceted and typically involve a mix of utility and private projects of various scales and at different stages of progress. Large-scale projects usually have high-reliability requirements and the availability of a skilled workforce. Infrastructure may be located throughout the grid, and network elements such as ring topologies (and associated protection) may be used to manage bidirectional power flows in the distribution network. As with medium-scale projects, progress toward high renewables is likely to align with reducing technology costs and increasing operational experience.

(vi)   Financing may be from multiple sources, including the private sector. Implementation is likely to require a mix of utility assets delivered through EPC-type contracts and IPPs operating under PPAs.

Popua Power Station, Tongatapu, Tonga. An example of a large-scale project (photo by TPL).

Cook Islands, Rarotonga Airport South
Load Shifting BESS (Photo by Entura).

Case Studies

Both the Outer Islands Renewable Energy Project (OIREP) and Tonga Renewable Energy Project (TREP) were driven by Tonga's Nationally Determined Contribution (NDC) targets of 50% renewable electricity by 2020 and 70% by 2030 (ARUP 2010). The Cook Islands Renewable Energy Sector Project (CIRESP) was driven by the Cook Islands NDC targets of 50% islands being powered by renewable electricity by 2015 and 100% by 2020 (Ministry of Finance and Economic Management 2017).[85]

The case study projects for Tonga and the Cook Islands were not, on their own, expected to deliver these NDCs fully but were established as crucial and substantial steps toward those goals. They were also intended to be compatible with and support future developments toward the respective targets.

---

[8]   The Cook Islands is currently finalizing revised targets for the 2020–2030 period.

3.1 Outer Islands Renewable Energy Project in Tonga

Case studies under OIREP are also grouped into three categories according to the scale—though there were only small and medium projects, but no large projects. These are presented in the following subproject descriptions (Table 3.1).

Table 3.1: Outer Islands Renewable Energy Project Profile

OIREP Features	Subproject: 1
Island Ha'apai	**Schedule:** 2014–2017
Island type Secondary population centre	**Grid scale:** Medium
Approximate budget $17 million	**Source of finance:** Asian Development Bank, Australian Department of Foreign Affairs and Trade, Global Environment Facility, Government of Tonga
Existing infrastructure	1 x 300 kW diesel generator 2 x 186 kW diesel generators 11 kV / 400 V distribution network
Financed project infrastructure	550 kW Solar PV 500 kW / 660 kWh BESS Controller
Associated project infrastructure	N/A
Site photo	 **Ha'apai OIREP solar PV array.** OIREP Subproject 1 (photo by Entura)
Objective	Increase renewable energy contribution on Ha'apai from 0% to 50%.
Project concept and evolution	This subproject was designed to provide fuel savings through solar PV generation. The 550 kW capacity of the solar PV plant proposed meant that solar PV output would regularly exceed load. Thus, a 660 kWh BESS was included to store excess power. The battery storage also incorporated isochronous grid-forming inverter capability, meaning diesel generators could be turned off when solar output is high, thus increasing diesel savings. A controller was required to manage the operation and integration of the plant.

continued on next page

Table 3.1 *continued*

OIREP Features	Subproject: 1
Project status	The project EPC contract was completed and commissioned in 2017. There were a number of initial technical challenges in commissioning (including earthing arrangements and operation of the control), which could be attributed to the relative novelty of this solution. Planned renewable energy percentage contributions have not been achieved (based on qualitative observations of the operators, as quantitative data was not available). However, this is attributable to significant load growth on Ha'apai, which means that while the renewables provide the expected energy, their contribution appears lower as a percentage of load. The target is expressed as a percentage of load. Considering the performance indicators in Appendix 2, the project is considered to have satisfactory technical performance. Although it did have major defects in the operation of the BESS control, these were resolved during the defects liability period (DLP). Renewable benefits in energy terms, performance warranties, and network impact are as planned. Project delivery performance was reduced due to delays in completion and commissioning. Nevertheless, the project was delivered in accordance with planning.

OIREP Features	Subproject: 2
Island 'Eua	**Schedule:** 2014–2017
Island type Secondary population center	**Grid scale:** Medium
Approximate budget $0.8 million	**Source of finance:** Asian Development Bank, Australian Department of Foreign Affairs and Trade, Global Environment Facility, Government of Tonga
Existing infrastructure	1 x 300 kW diesel generator 2 x 186 kW diesel generators 11 kV / 400 V distribution network
Financed project infrastructure	200 kW Solar PV Controller
Associated project infrastructure	N/A
Site photo	 **'Eua OIREP solar PV array.** OIREP Subproject 2 (photo by Entura).
Objective	Increase renewable energy contribution on 'Eua from 0% to 17%.

continued on next page

Table 3.1 *continued*

OIREP Features	Subproject: 2
Project concept and evolution	This subproject was designed to provide fuel savings through solar PV generation. The 200 kW capacity of the solar PV plant proposed meant that solar PV output would sometimes exceed load. Thus, a controller was added that measured load and diesel generation minimum output, and curtailed the solar PV as needed to avoid oversupply, which would have caused a blackout. The controller was also able to schedule which diesel generator was online, according to load and available solar resources.
Project status	The project EPC contract was completed in 2017. As per the performance indicators in Appendix 2: there were no major defects, performance warranties were met, and network impact was substantial as planned. However, the larger diesel generators were manually operated to ensure grid stability under rapidly fluctuating solar PV, which was not fully managed by the installed controller and resulted in reduced fuel savings relative to plan. Automatic control and storage implemented under TREP are intended to resolve this problem. Project delivery performance was considered acceptable, with only minor delays during implementation and commissioning.

OIREP Features	Subproject: 3
Islands Niuafo'ou, Niuatoputapu, 'Uiha, Nomuka, Ha'ano, Ha'afeva	**Schedule:** 2014–2017
Island type Secondary population centre	**Grid scale:** Small
Approximate budget $3.6 million	**Source of finance:** Asian Development Bank, Australian Department of Foreign Affairs and Trade, Global Environment Facility, Government of Tonga
Existing infrastructure	Niuafo'ou and Niuatoputapu: Some solar home systems 'Uiha, Nomuka, Ha'ano, Ha'afeva: Diesel mini-grids
Financed project infrastructure	Niuafo'ou: 183 kW solar home systems Niuatoputapu: 150 Solar PV, 295 kWh BESS and 80 kW backup diesel generator 'Uiha: 100 kW Solar PV, 210 kWh BESS and 50 kW backup diesel generator Nomuka: 100 kW Solar PV, 210 kWh BESS and 50 kW backup diesel generator Ha'ano: 100 kW Solar PV, 210 kWh BESS and 50 kW backup diesel generator Ha'afeva: 60 kW Solar PV, 110 kWh BESS and 30 kW backup diesel generator
Associated project infrastructure	N/A
Site photo	 **Site preparation at Ha'ano.** OIREP Subproject 3 (photo by Ministry of Meteorology, Energy, Information, Disaster Management, Environment, Climate Change and Communications).

continued on next page

Table *continued*

OIREP Features	Subproject: 3
Objective	Increase renewable energy contribution on each island from 0% to >90%.
Project concept and evolution	This project was designed to improve energy access and add renewable energy to existing diesel mini-grids on selected outer islands.
Project status	The project EPC contract was commenced in 2017. To date, the replacement diesel generators have been commissioned. However, solar PV and BESS is currently awaiting completion and commissioning. Consequently, technical performance cannot be assessed. Project delivery performance, however, has been below expectations, primarily due to delivery delays and various minor technical issues. This is frequently attributed, by stakeholders, to a lack of in-country expert management of the project delivery.

BESS = battery energy storage system, EPC = engineering, procurement, and construction, kW = kilowatt, OIREP = Outer Islands Renewable Energy Project Profile, PV = photovoltaic, TREP = Tonga Renewable Energy Project.
Source: Author.

3.2 Tonga Renewable Energy Project

Case studies under TREP were grouped into three categories according to the scale (small, medium, large). These are presented in the following subproject descriptions (Table 3.2).

Table 3.2: Tonga Renewable Energy Project Profile

TREP Features	Subproject: 1
Island Tongatapu	**Schedule:** 2017–2022
Island type Main island	**Grid scale:** Large
Approximate budget $32.2 million	**Source of finance:** Green Climate Fund, Asian Development Bank, Australian Department of Foreign Affairs and Trade, Government of Tonga, Independent Power Producers
Existing infrastructure	16 MW diesel generation 1 MW solar PV 2 MW solar PV 1.3 MW solar PV (TPL) 11 kV / 400 V distribution network
Financed project infrastructure	Grid stability BESS (5.1 MW/2.5 MWh) Load-shifting BESS (5 MW/17.4 MWh)
Associated project infrastructure	3 x 2 MW solar PV 3.8 MW wind 2 MW wind
Site photo	**Grid stability BESS at Popua Power Station site.** TREP Subproject 1 (photo by TPL).

Table 3.2 *continued*

TREP Features	Subproject: 1
	Load shifting BESS. TREP Subproject 1 (photo by TPL).
Objective	Increase renewable energy contribution on Tongatapu from 15% to 54%.
Project concept and evolution	This subproject was originally conceived as distributed solar or wind battery installations into the existing grid, with no centralized storage. Preliminary technical analysis under the feasibility stage demonstrated that a centralized battery concept for grid stabilization (first priority) and energy storage (second priority) could offer higher reliability, lower complexity, and better return on investment. Both grid stability and load-shifting BESS could support a range of additional distributed renewable generators in the network (beyond the associated generation identified for the project). Consequently, the subproject concept became one of 2 x centralized BESS.
Project status	The grid stability BESS completed commissioning in December 2021. The load-shifting BESS is expected to complete commissioning in January 2022. Commissioning of the first associated generation (2 MW solar PV) is complete. Technical performance has not been evaluated as the project is not yet operational. Project delivery performance has been hampered by COVID-19-related travel restrictions; otherwise, no significant issues were observed.

TREP Features	Subproject: 2	
Islands Vava'u, 'Eua	**Schedule:** 2017–2023	
Island type Secondary population center	**Grid scale:** Medium	
Approximate budget $5.1 million	**Source of finance:** Green Climate Fund, Asian Development Bank, Australian Department of Foreign Affairs and Trade, Government of Tonga	
Existing infrastructure	Vava'u: 2 MW diesel generation 420 kW Solar PV 117 kWh BESS 11 kV / 400 V distribution network	'Eua: 3 x diesel generators 200 kW solar PV 11 kV / 400 V distribution network
Financed project infrastructure	Vava'u: 300 kW Solar PV Grid stability BESS (0.9 MW/0.45 MWh)	'Eua: 350 kW Solar PV Grid stability and load-shifting BESS (0.4 MW/0.9 MWh)
Associated infrastructure	N/A	

continued on next page

Table 3.2 *continued*

TREP Features	Subproject: 2
Site photo	 **'Eua existing infrastructure.** TREP subproject 2 (photo by Entura).
Objective	Increase renewable energy contribution on Vava'u and 'Eua from 12% to 22%.
Project concept and evolution	This subproject was originally conceived as the addition of solar and BESS at each island to boost renewable contribution. For Vava'u, the proposed solar was, however, at a threshold where grid stability was only marginally compromised. An option was proposed to forgo BESS and instead curtail solar when required. However, this option was not selected due to the potential risk to supply. As a consequence, the selected BESS is expected to be very lightly cycled with the current solar PV capacity. The scope to increase the solar PV capacity (up to about 1 MW on the current site) without additional BESS is under consideration.
Project status	The project EPC contract was awarded in March 2020. Design and procurement is complete. However, construction has been delayed due to COVID-19 travel restrictions. Subject to further COVID-19 related delays, completion is planned for the end of 2022.

continued on next page

Table 3.2 *continued*

TREP Features	Subproject: 3
Islands Niuafo'ou, Kotu, Tungua, O'ua, Mo'unga'one	**Schedule:** 2017–2023
Island type Outer islands	**Grid scale:** Small
Approximate budget $10.9 million	**Source of finance:** Green Climate Fund, Asian Development Bank, Australian Department of Foreign Affairs and Trade, Government of Tonga
Existing infrastructure	Some existing solar home systems (solar panel and battery). A small number of privately owned portable generators.
Financed project infrastructure	Niuafo'ou: 250 kW Solar PV, 126 kW/2,2755 kWh BESS Kotu: 70 kW Solar PV, 36 kW/580 kWh BESS Tungua: 70 kW Solar PV, 36 kW/590 kWh BESS O'ua: 60 kW Solar PV, 36 kW/470 kWh BESS Mo'unga'one: 50 kW Solar PV, 27 kW/390 kWh BESS
Associated project infrastructure	N/A
Site photo	 **Proposed Hybrid Power Station Site at Kotu, Ha'apai.** TREP subproject 3, typical solar / power station site, precommencement (photo by Entura).
Objective	Electrification for 100% of the population from a baseline of 0%. Electrification via 100% renewable energy.
Project concept and evolution	This subproject was originally conceived as solar and BESS mini-grids with diesel backup. Due to the requirement of one funder (GCF), diesel was removed from the project definition. Requirements for an equivalent high level of reliability (e.g., 5 days autonomies) were set by the implementing agency (IA) and resulted in a substantial increase in BESS size. These projects, on their own, were not able to provide full capital cost recovery and instead were required to be financially sustainable— with tariffs set to cover O&M only.
Project status	The project EPC contract was awarded in March 2020. Design and procurement are complete. However, construction has been delayed due to COVID-19 travel restrictions. Construction planning will rely on detailed site visits, which cannot commence until COVID-19 travel restrictions are lifted. Completion of the projects is therefore expected in early 2023.

BESS = battery energy storage system, MW = megawatt, MWh = megawatt-hour, O&M = operation and maintenance, PV = photovoltaic, TREP = Tonga Renewable Energy Project.
Source: Author.

3.3 Cook Islands Renewable Energy Sector Project

Case studies under CIRESP are also grouped into three categories according to scale (small, medium, large). These are presented in the following subproject descriptions (Table 3.3).

Table 3.3: Cook Islands Renewable Energy Sector Project Profile

CIRESP Features	Subproject: 1
Island Rarotonga	**Schedule:** 2016-2022
Island type Main island	**Grid scale:** Large
Approximate budget $32M	**Source of finance:** GCF, GEF, ADB, GCI, IPP
Financed project infrastructure	Grid stability BESS (6 MW/3 MWh) Load shifting BESS (2 MW/8 MWh) Load shifting BESS (1 MW/4 MWh)
Associated project infrastructure	At least 6 MW solar PV—expected to be installed by the private sector
Site photo	 Airport South load-shifting BESS on Rarotonga, Cook Islands under construction. CIRESP subproject 1 (2 MW/8 MWh) (photo by Entura). Airport West load-shifting BESS on Rarotonga, Cook Islands. CIRESP subproject 1 (1 MW/4 MWh) (photo by Entura).

continued on next page

Table 3.3 *continued*

CIRESP Features	Subproject: 1
Objective	Increase renewable energy contribution on Rarotonga from 15% to 39%.
Project concept and evolution	The first load-shifting BESS was initiated because grid stability limits were preventing increased private sector take-up of solar PV. A load-shifting BESS was selected as the utility was not yet prepared for a grid stability BESS, and because at the time, grid stability BESS were highly complex to implement. This BESS (1 MW/4 MWh) had demonstrated financial feasibility and the potential to support the addition of 2 MW of solar PV at the cost of approximately $4 million. Subsequently, a need was identified for increased grid support on Rarotonga to absorb intermittent electricity to be generated by privately financed solar PV and wind power plants. At the same time, a GCF grant funding opportunity to support renewable energy in the Cook Islands was available as part of a program initiated by ADB for the Pacific (Pacific Renewable Energy Investment Facility [PREIF]). The indicative value of this grant was $12 million, which would allow three times the energy storage of the first BESS, and therefore 6 MW of additional solar PV. Prior to procurement and through discussions with the utility and technical advisors, it was identified that the benefits of the first BESS (1 MW/4 MWh) were not directly scalable, and a higher power capacity was necessary for grid stability under increased renewable penetration. Therefore, the scope of the GCF grants-funded BESS was adjusted to increase power and reduce the storage of one unit of BESS, keep to the approved budget and provide nominally similar BESS specifications. That is, instead of three units of 1 MW/4 MWh BESS, there were two units of 1 MW/4 MWh "load-shifting" BESS and one unit of "grid stability" 4 MW/1 MWh BESS.[a]
Project status	Both load-shifting BESS is complete and operational (completed September 2019 and February 2020). However, neither BESS is significantly loaded as only approximately 1 MW of the planned >6 MW total associated generation has been completed. Both load-shifting BESS satisfied performance warranty requirements. However, their broader impact on the network is yet to be quantified. One of the two BESS experienced significant delivery issues due to insufficient resourcing and complex, bespoke control, and integration design. These have since been resolved, though some quality-related defects remain open pending resumption of travel to the Cook Islands. The grid stability BESS is due for commissioning in February 2022 (having experienced delays due to COVID-19 travel restrictions). Associated generation was initially added through pending connections of BTM solar PV (approximately 1 MW solar PV). However, the further installation has been put on hold pending completion of the power station BESS, control system, and tariff arrangements.

ADB = Asian Development Bank, BESS = battery energy storage system, CIRESP = Cook Islands Renewable Energy Sector Project, EU = European Union, GCI = Government of Cook Islands, GCF = Green Climate Fund, GEF = Global Environment Facility, IPP = independent power producer, MW = megawatt, MWh = megawatt-hour, PV = photovoltaic.

[a] The MW rating here was the overload capability for short-term frequency response. Continuous power rating was 50% of the overload capability.

Table 3.3 *continued*

CIRESP Features	Subproject: 2
Island Aitutaki	**Schedule:** 2017-2019
Island type Secondary population centre	**Grid scale:** Medium
Approximate budget $3 million	**Source of finance:** ADB, GCI
Existing infrastructure	3 x 600 kW diesel generation 6.6 kV and 11 kV / 400 V distribution network
Financed project infrastructure	Grid stability BESS (0.5 MW/0.5 MWh) 750 kW Solar PV 300 kW diesel generator Control system Load shifting BESS (1 MW/4 MWh)
Associated project infrastructure	N/A
Site photo	 **Grid stability BESS on Aitutaki, Cook Islands.** CIRESP subproject 2 (0.5 MW/0.5 MWh) (photo by Entura).
Objective	Increase renewable energy contribution on Aitutaki from 0% to 24%.
Project concept and evolution	The initial concept for Aitutaki was to install a small (300 kW) diesel generator, solar PV (750 kW) control system. The purpose of the small diesel generator was to provide increased operational flexibility in scheduling of generators and to have a smaller generator operating (with a lower minimum load) when solar PV output was high, to maximize the use of solar energy. A BESS was initially excluded due to cost. Due to concerns about the reliability of forecasting and control as a means to schedule diesel generators under varying solar PV output, a BESS was included to provide a buffer for supporting the small diesel generator if solar output dropped while a larger generator was started. The revised project concept was acceptable to stakeholders and implemented on that basis.
Project status	The Aitutaki project was completed in 2019 and has performed to expectations, including on-time delivery to specification, the satisfaction of performance warranties, expected network impact, minimal defects resolved during DLP, and matching of planned renewable energy contribution in both absolute (960 MWh) and percentage terms (24%). This system is currently saving over 300,000 liters of fuel annually.

continued on next page

Table 3.3 *continued*

CIRESP Features	Subproject: 3
Islands Atiu, Mauke, Mangaia, Mitiaro	**Schedule:** 2015–2019
Island type Outer island	**Grid scale:** Small
Approximate budget $9million	**Source of finance:** ADB, EU, GCI
Existing infrastructure	1–3 diesel generators per island (<100 kW each) 400 V or 6.6 kV / 400 V distribution network
Financed project infrastructure	Mitiaro: 159 kW solar PV, 72 kW/972 kWh BESSa Mauke: 229 kW solar PV, 90 kW/1214 kWh BESS Atiu: 413 kW solar PV, 162 kW/2186 kWh BESS Mangaia: 477 kW solar PV, 216 kW/2915 kWh BESS
Associated project infrastructure	N/A
Site photo	**Mitiaro power house opening, Cook Islands.** CIRESP subproject 3 (photo by Entura).
Objective	Increase renewable energy contribution on each island from 0% to 90%–95%
Project concept and evolution	The project concept for these islands was largely consistent with earlier projects undertaken in the Northern Group of the Cook Islands. For this reason, equipment compatibility was sought, and lead-acid batteries were selected. There was considerable discussion at the time about lithium-ion alternatives. However, in 2015, lithium-ion BESS were substantially more expensive (even allowing for higher power density), required active temperature control (which could not be guaranteed on the outer islands), and posed a moderate fire risk (also unacceptable on outer islands).
Project status	All projects were completed in 2019. Their performance varies in terms of defects and renewable energy contribution. As for Ha'apai, this has been impacted in some instances by load growth that has been above projections. However, generally approximately 90% fuel savings have been achieved against a 90%–95% target. But for the island of Mangaia, the largest island and the one experiencing the most underperformance (68% renewable energy compared to 92% planned), the most significant issue has been integration with the existing diesel generators.
	These generators experienced frequent faults when operating with the renewable energy power system. In response, the operators would anticipate problematic conditions and run diesel-only during these times, reducing the renewable energy contribution coinciding with the end of their useful operational life. CIG plans to resolve this issue by upgrading the diesel generators within the next 2 years.

[a] Each BESS is sealed lead-acid technology, rather than lithium-ion used for other projects. Thus, energy capacity is not directly comparable. Minimum state of charge for sealed lead-acid batteries was approximately 50% compared to 5% for lithium-ion ones.

Source: Author.

3.4 General Comments on Project Concept and Concept Evolution

In several cases, the final design of the case studies differed from the original concept. The typical process for formulation and finalization of subproject concept was as follows:

(i) Early concepts were developed in-country between government and utility representatives in response to NDC and policy targets. In some cases, these concepts were based on consultant technical studies, but concepts or preferences were also directly proposed by the government (political representatives or bureaucrats) or utility representatives.

(ii) Consultation occurred between government stakeholders and funding agencies to communicate high-level country priorities and ascertain funding availability. Inevitably, funding would not be sufficient to progress all subproject concepts, and subprojects that progressed through this process would typically be driven by a combination of internal lobbying of stakeholder interests, value for money, and negotiation against funding agency priorities. Preliminary safeguards and financial viability screening would be performed at this stage.

(iii) The resulting subproject concepts would then be the basis for securing funding, which could be unique in other regions but common in the Pacific due to the limited funding envelope.

(iv) In-depth technical assessment, optimization of the concept, and detailed financial, economic, and safeguards checks were then undertaken as part of the final feasibility and/or due diligence phase. Any technical adjustments and optimization were made within the scope of the basis for securing funding (the projects were substantially similar to the description of the funded project). Project feasibility was demonstrated prior to contracting.

3.5 General Comments on Project Status

In almost all cases, projects took longer than initially scheduled. In one case with the longest delay, the project was delivered approximately 3 years later than estimated during planning. The reasons for delays were varied and included the following:

(i) Factors typical of any infrastructure project, such as administrative or procurement procedural issues (these included retendering due to no bidders meeting qualification or technical response requirements; small increments in time for review of bidding documents, tender evaluation, approval of award; delays in completing contract preliminaries including advance guarantees and letter of credit; discrepancies in the information provided for invoicing or registrations; legal reviews and resolution of country-specific regulations inconsistent with ADB procurement rules).

(ii) Contractor delivery delays (resulting in delay liquidated damages), attributed to undercommitment of contractor resources to the project.

(iii) COVID-19 delays were affecting factory test, equipment shipping to a site, and travel of key personnel to site.

The second and third issues will be considered further in the project analysis in section 6, as some of these derive from the unique geographic isolation of Pacific SIDS. The general implication that most projects run behind schedule (irrespective of the cause) will also be considered in terms of its program impact.

In terms of project performance, those projects that are completed showed varying technical performance. In the majority of cases, basic requirements were met with specifications met, defects resolved during DLP, performance warranties met and network impact as projected. However, in several cases there were extended issues with technical performance observed. The rate of such projects (approximately 40%) was high, reflecting the challenges inherent in these projects, which are addressed in section 6.

Additionally, even where projects performed to expectations, targets expressed as a percentage renewable energy contribution proved difficult to meet since these targets were constantly changing with demand. Expression of targets in terms of absolute energy contribution would significantly simplify performance reporting.

Tonga, Ha'apai small scale solar PV and BESS
(Photo by TPL).

4

Scope and Methodology of Review

This review provides an in-depth analysis of BESS and BESS hybrid project implementation. It focuses on implementation-specific project stages and requirements of the selected case studies. There is significant further scope for exploring related project stages and requirements of these case studies (such as financing and operations) suitable for subsequent in-depth analysis, which is not covered here.

This section provides a brief overview of the project stages and requirements covered as part of the implementation, as well as a description of the selected case studies.

4.1 Project Challenges

4.1.1 Project Life Cycle

The following is a high-level list of all the typical stages of implementation of a BESS or BESS-hybrid energy project in isolated grids. This is presented in sequential order, with those stages particularly relevant to implementation, and the subject of the analysis in this report is highlighted in green.

Figure 4.1: Project Stages—Main Implementation Stages Shown in Green

	Early	Year 1	Year 2	Year 3	Year 4	Year 5	Future
• Initiate							
o Need	█						
o *Concept*	█						
o *Initial subproject selection*		█					
o Develop support		█					
• Planning							
o Financing and budget		█					
o Implementation structure		█					
o Technical assistance			█				
o Management Structure			█				
o *Schedule*			█				
o *Feasibility*			█				
o Investment decision				█			
• *Executing*							
o *Specification*			█				
o *Procurement*			█				
o *Technical oversight*				█			
o *Contract management*				█			
o *Design*				█			
o *Logistics*				█			
o *Construction*					█		
o *Documentation*					█		
o *Test and Commissioning*					█		
o *Training*					█		
• Operations							
o *Warranties*						█	
o *Defects*						█	
o *Hand-over*						█	
o *Spare parts*						█	
o Performance monitoring						█	
o Routine operations						█	
o Fault response						█	
o Data management						█	
o Long term maintenance						█	
• Closing							
o Decommissioning, disposal and recycling							█
o Replacement or refurbishment							█

Note: Main implementation stages shown in green.
Source: Author.

4.1.2 Project Requirements

Project requirements are considered here as a separate dimension to the project stages. They relate to the knowledge, capability, technology, tools, and procedures needed for the project to work effectively.

Technical

Technical project requirements pertain to the capability, maturity, and uncertainty of hardware, data, models, and industry sectors to meet the project needs. For BESS and BESS hybrid projects, the following are considered the core technical requirements for implementation.

(i)   Core technology options

    (a)   Selection of the likely mix of technologies considering the existing project infrastructure, renewable resources, land availability, and project objectives. Technologies are typically selected from energy storage (BESS or pumped hydropower); renewable generation (solar PV, wind power, mini-hydro, biomass, etc.).

    (b)   Energy balance modelling is typically applied to determine the optimal mix of capacities from the selected technology.

(ii)   Demand profile and growth

    (a)   Energy balance modelling also requires projections of estimated hourly demand over the project life. Typically sensitivity studies are included given the uncertainty in growth projections for many locations.

(iii) Reliability requirements

    (a)   It is important to quantify the reliability levels required for the project. Where possible, standard metrics related to customer interruptions are used (e.g., System Average Interruption Duration Index [SAIDI]) in conjunction with generator and transmission redundancy levels and reserve requirements for variability in renewable energy resources.

    (b)   The reliability requirements are used to analyze various scenarios in energy balance and power systems modelling.

(iv)   Civil and electrical balance of plant

    (a)   A broad range of requirements must be considered, such as soil and hydrological conditions, standards for survivability, resistance to local environmental conditions and pests, and general adherence to local and international codes and standards.

(v)   Control and integration

    (a)   Developing a functional specification for control and integration of various technical elements of the network is a fundamental requirement for high renewable energy contributions. The design of control systems and communication interfaces follows the functional specification and typically adopts a hierarchical approach, with high-level dispatch control relying on plant-level controller performance.

(vi)   Power systems modelling

    (a)   For more complex systems, power system modelling is necessary to understand the response of the network, generation, and protection elements to critical events. Critical events are selected from energy balance modelling and reliability requirements. Where necessary, special requirements for control system performance may be specified based on power system modelling.

    (b)   For simpler systems, behavior may be predictable based on standard equipment capabilities and power systems modelling is not required.

(vii) Safety, environment and standards

    (a)   Monitoring and implementing the most up-to-date requirements for safety and environment, as well as performance and testing, are critical to maintaining long-term sustainability for each project and confidence in the broader program.

Procurement and Contract Management

Procurement and contracting arrangements for each of the case studies were undertaken following ADB's Procurement Guidelines (2015, as amended from time to time) and consulting services in accordance with ADB Guidelines on the Use of Consultants (2013, as amended from time to time), as well as local procurement policies. Where discrepancies arose, ADB policies took precedence and were a requirement of the Financing agreement. Therefore, this section and the report generally focus on applying ADB procurement processes and standard contracts to the project, noting that each tender still required approval from local procurement agencies, which impacted schedule.

Procurement and contracting requirements are very broad, and many aspects are standardized and consistent with industry practice. The scope of ADB requirements is well documented.[9] However, the following aspects were considered of particular importance for the case studies considered here:

(i)   Procurement packaging: division of project requirements amongst separate contracts

(ii)   Standard bidding documents: suitability of available procurement methods for innovative projects

(iii)   Tender evaluation criteria: quantifying qualitative factors in the least cost procurement process

(iv)   Compatibility of qualification criteria with country regulations and typical structuring of contractors: how to treat special purpose vehicle companies, and qualifications of parent companies and subsidiaries.

(v)   Liquidated damages, payment milestones and performance guarantees: extent of contract manager's control over contractor delivery.

(vi)   Insurance: insurance options for a maturing product in a niche market.

Stakeholder engagement

There is a wide range of stakeholders in all the case studies:

(i)   Government political leaders (central and island councils)

(ii)   Donors and other financiers

(iii)   Government bureaucrats, including RE program developers, regulators, and donor support agencies

(iv)   Electricity utility or relevant authority (and asset owner/manager if separate), and individual plant operators

(v)   Landowners

(vi)   Contractors and O&M providers, and their suppliers

(vii) Customers and communities generally.

The role of the various stakeholders at different project stages will be examined in the scope of this assessment. In particular, the relative role in decision-making and the impact of decisions on stakeholders will be a core focus.

4.1.3 COVID-19 Impact

Implementation of most case studies included the period 2020–2021 and thus were significantly impacted by COVID-19. It is inevitable, therefore, that the scope of this assessment must also consider such impacts.

---

[9]   ADB. 2017. *Procurement Regulations for ADB Borrowers*. Manila.

4.2 Review Methodology

The methodology used in preparing the assessment for this report is adapted from a standard project post-implementation review (PIR) and applied across a portfolio of projects (Briscoe et al. 2000). The lead reviewers, in this case, are the paper's authors, and the subject matter experts relied upon are the core project delivery team and key stakeholder representatives.

The focus is on steps 4 and 5 of the PIR, e.g., (4) Findings: a summary of the issues found during the review process; and (5) Recommendations: actions to be taken to correct findings. Earlier stages of project review, including compliance and project metrics, have been completed during the course of the project (and are documented in the project information progress reporting). The objective is to reflect on the experiences of those delivering the project, with an open perspective on how to provide improvement for future opportunities.

The advantage of this approach is the ability to reflect a very deep understanding of project implementation issues and successes from a range of perspectives. The main disadvantage is the potential to overlook potentially important factors that an independent insight or project audit may reveal. On balance, considering the different perspectives and experiences of the delivery team (including outside the case studies), the PIR process was considered warranted. The authors also note that there has previously been some level of independent scrutiny at the project level and the delivery team initiated by the Government of the Cook Islands. Limited informal feedback was provided on areas for improvement to the delivery team, but no significant concerns or noncompliances were raised.

4.3 Project Documentation

The analysis included a review of documents in the public domain, including feasibility studies, project administration manual, and project progress reporting (Asian Development Bank 2021, Ministry of Finance and Economic Management, 2021). The analysis also included documents available to the authors but with limited circulation or commercially sensitive—any information derived from such documents is amalgamated or has sensitive details removed.

4.4 Stakeholder Consultation

Table 4.1 lists the stakeholders interviewed for this assessment.

Table 4.1: Stakeholders Consulted

Country	Project	Name	Organization	Position	Organisation type
Tonga	TREP	Finau Katoanga	TPL	Project Manager	Utility
Tonga	OIREP and TREP	Setitaia Chen	TPL	CEO (former)	Utility
Tonga	TREP	Nikolasi Fonua	TPL	Engineering Manager, Acting CEO (current)	Utility
Tonga	OIREP and TREP	Michael Lani 'Ahokava and Murray Sheerin	TPL	Power Station Managers	Utility
Tonga	OIREP and TREP	Simon Wilson	TPL / PMU	Major Project Manager / Project Manager	Utility / PMU
Tonga	TREP	Adrien Bock	Akuo Energy	Business Development Manager	EPC contractor
Tonga	OIREP	Ajay Prasad	AUSPAC energy	Business Development Manager	EPC contractor
Cook Islands	CIRESP	Tangi Tereapii	REDD	Director	Government
Cook Islands	CIRESP	Lesley Katoa	TAU	CEO (current)	Utility
Cook Islands	CIRESP	Apii Timoti	TAU	CEO (former)	Utility
Cook Islands	CIRESP	Tei Nia	TAU	Active Chief Engineer	Utility
Cook Islands	CIRESP	Long Tuiravakai	TMU	Power Station Manager	Utility
Cook Islands	CIRESP	Romani Katoa	PMU	Project Manager	PMU
Cook Islands	CIRESP	Anthony Whyte	Mangaia	Executive Officer	Island Council
Cook Islands	CIRESP	Ben Tavai	MFEM	Client representative	Government
Cook Islands	CIRESP	Steve Anderson	Andersons	Director	Supplier
Cook Islands	CIRESP	Dean Parchomchuk	Vector-Powersmart	Project Manager	EPC contractor
All	All	David Skinner	Entura	Renewable Energy Engineer	Technical Assistance

CEO = chief executive officer, CIRESP = Cook Islands Renewable Energy Sector Project, OIREP = (Tonga) Outer Island Renewable Energy Project, PMU = Project management unit, REDD = Renewable Energy Development Division, TAU = Te Aponga Uira.
Source: Author.

Initially, stakeholders were asked a series of general questions to identify focal points for this assessment. These questions, and the summary responses, are included in Appendix 1.

Where it was relevant to understand the focal areas better, follow up questions were asked of the same stakeholders. Responses are included where relevant in section 5.

4.5 Literature Review

A literature review was undertaken to identify existing information related to best practices for

(i)   key BESS and hybrid system technology and standards;

(ii)   procurement of BESS and similar novel, maturing or innovative technologies; and

(iii)   stakeholder engagement and decision-making in the energy sector, focusing on SIDS and DMS.

The intent was to understand whether the findings identified were consistent, inconsistent, or additional to existing information on these areas; and if they employ existing information to understand better or interpret the findings of this assessment.

4.5.1 BESS and Hybrid System Technology and Standards

BESS Capabilities

Over the past years, battery systems have developed to perform their traditional storage role and provide network support functionalities to increase the system's resilience to faults and disturbances. Fast frequency response, virtual inertia, and grid-forming capabilities are some of the tools that have recently been developed and added to the grid-connected inverters used in the batteries currently available in the market (Chaudhary et al. 2021; Pattabiraman, Lasseter, and Jahns 2018; Cherevatskiy et al. 2020; Lin et al. 2020).

Inverter based resources displace conventional synchronous generation (Pattabiraman, Lasseter and Jahns 2018). Whereas conventional synchronous generation provides significant physical inertia, inverter based resources do not. Thus, hybrid systems introduce net reduction in system inertia and, without any other changes, this can result in a corresponding increase in frequency excursions during normal operations and contingency events.

Fast frequency response entails the integration of a low-level control loop within the BESS inverter that varies real power as a function of frequency (typically with a deadband). The integration of this control loop rather than requiring external set-point control means BESS can typically respond to frequency deviations with a full-scale change in power output in less than 50 milliseconds (compared to 2 seconds or more for external setpoint).

This is comparable to or better than the rise time due to the inertia of synchronous diesel generators.

Shaping of the fast frequency response through the control algorithm is also possible, allowing the BESS response to closely mirror the shape of the response due to the inertia of a synchronous diesel generator. Various proprietary algorithms are implemented to deliver this effect, intended to provide a "virtual synchronous machine" that simplifies integration and operation of existing network systems (such as protection).

However, even with fast frequency response, grid-following (or grid-tied) inverters, typical of solar inverters and conventional BESS inverters are still dependent on synchronous generation to 'form the grid'. That is, to provide a reference voltage source waveform. Newer generation grid-forming inverters act as a voltage source and can generate their own reference, which can be modulated relative to the grid isochronous source (even at the sub- cycle level) using fast frequency control, or which can act as an independent isochronous generator— forming a grid with no synchronous generation.

The development of the grid forming feature has allowed hybrid systems to operate at higher penetrations of renewable energy or weak grids since solar PV, wind, and most alternative generation sources require a reference voltage and frequency source. It is worth noting that inverters have not only developed grid-forming capabilities but the ability to operate at very low short circuit ratios, significantly beyond what conventional inverter-based generation can perform. These elements have significantly pushed the boundaries and limitations of renewable contribution, especially in isolated hybrid power systems.

A project implemented in Australia at Dalrymple (Leitch, 2020) demonstrates the capability to employ grid forming BESS and virtual inertia to support weak, isolated grids. The extensive knowledge base from this project supports the application of BESS, with appropriate control and fault current capability. In this instance, the size of the BESS is substantially larger than the local peak load (sizing considers applications when the grid is not isolated and/or islanded), allowing the provision of all network services.

Standards for BESS

In the early stages of BESS procurement, the lack of defined performance levels and technical requirements and standards meant comprehensive detail was included in specifications. This created risk (of error or omission) and also resulted in challenges from manufacturers who had their own interpretation and assumptions with regard to items such as measuring energy storage, efficiency, ramp rate or response time requirements; definition of terms; access provisions; grid support functions; duty cycle; end of life; warranty limits, and more. This led to misalignments between technical specifications and the products offered by manufacturers. In the absence of standards, the abovementioned performance requirements had to be specified with the best knowledge of technical personnel with non-standardized details and requirements to set minimum quality parameters expected.

Critically, a lack of clear standards and operational guidelines were partially responsible for safety incidents, including a series of fires in BESS over the past few years, such as those due to battery protection system failure in the Republic of Korea (Hering 2019).

These challenges were not unique to isolated grids, however, and were well recognized by the broader industry. Newly established committees have developed a multitude of recently published guidelines, codes, and standards for battery energy storage systems during the last few years. These are available to help streamline and standardize the process of safely and effectively deploying BESS. Key publications include the following:

(i) IEC 62933: Electrical energy storage (EES) systems, including

  (a) definition of BESS, BESS equipment, configuration, performance, and tests
  (b) guidance on BESS application, selection and implementation
  (c) comprehensive safety requirements, including specific sections suited for lithium-ion BESS
  (d) initial requirements for environmental risk assessment

(ii) UL 9450A Test method: Testing the fire safety hazards associated with propagating thermal runaway within battery systems.

(iii) IEC 62902(2019) Secondary cells and batteries: Marking symbols for identification of their chemistry

(iv) IEC 62281: 2019 Safety of primary and secondary lithium cells and batteries during transport

This is only a small subset. DNV-GL (2019) provides a more comprehensive gap analysis of existing standards (noting further developments since that time) related to BESS performance. This analysis concluded that while there were a range of different standards (over 124 reviewed) for the performance of various BESS chemistries and components, published by a range of reputable international standards organizations. There was a lack of an overarching systems level standardization for performance. Different manufacturers may select different standards to report against, resulting in inconsistency and complexity in comparing and evaluating benefits. Some of the standards mentioned earlier, particularly IEC 62933, which has advanced significantly since the DNV-GL analysis, are targeted at addressing such gaps.

4.5.2 Procurement of BESS and Similar Novel, Maturing, or Innovative Technologies

There is an extensive analysis of different procurement methods available to the energy and construction industry, for example, see Wardani et al. (2006) and Muriro and Wood (2010). The intent is not to revisit this here but rather to utilize such analysis to focus on the particular characteristics of hybrid electricity and BESS projects in isolated grids, which include

(i)   Low competition and/or small market

(ii)   Complex and bespoke projects, frequently integrated deep into existing systems

(iii)   Requiring innovation and use of emerging technology

(iv)   Robust and high-quality solutions

(v)   Frequently subject to external funding conditions with a focus on transparency and integrity; and

(vi)   Low risk to employer.

George and Egbu (2016) provide a framework for selecting procurement models based on such characteristics. This approach highlights the benefits of a partnership approach, which is well suited to complex, innovative projects and high-quality solutions. While not stated, this approach is likely to also suit a small market where traditional competitive options are likely to be limiting. However, the partnership approach does not rank as well for low risk to employer, where traditional design-build type solutions like EPC are preferred.

For similar projects with industrial clients, the authors have noted applications of early contractor involvement (ECI) contracting models. The ECI model provides an initial collaborative phase between the contractor and employer to advance preliminary design details, resolve risk elements, and enable a more refined basis of design and pricing structure compared to an EPC type contract. The use of this model is also supported by the literature (State Government of Victoria, 2021) and consistent with George and Egbu (2016) where uncertainty exists, or innovation is required in the project since it delivers a blend of partnership and traditional procurement models. The suitability of this approach is perhaps most strongly subject to the remaining driver above—being the external funding requirements. Such requirements are typically based on a generalized approach consistent with large, mainstream infrastructure projects (construction of a road, bridge, dam, conventional power station, port, etc.). Options to adapt such requirements to the project needs are reviewed in section 5.3.

Tonga Power Limited trained lines-workers
(Photo by TPL).

Principal Challenges for BESS and Hybrid Projects in Isolated Grids

Based on an analysis of stakeholder responses and other input data, the principal challenges outlined in this section were identified. These represented issues that

(i)   affected multiple projects;

(ii)   were raised by multiple respondents;

(iii)   were a structural or systemic issue, specific to this project type;

(iv)   were significant in terms of their impact;

(v)   had potential future opportunity or risk; and

(vi)   were relevant for the project implementation stage.

5.1   Technical Challenges

5.1.1   Technology: Maturing BESS Industry

At the commencement of the case study projects, BESS technology was in its infancy. While lead-acid battery variants were mature, these were rarely employed at the MW scale. BESS specific inverters were designed for home systems or microgrids or used generic variable speed drive type technology for bidirectional power flow, with custom software and control implementation. Most products were bespoke installations of various equipment, with limited integrated type testing or certification. BESS specific standards were lacking, and standards for other equipment or installations were adapted to the purpose.

Additionally, very few contractors had demonstrated experience with BESS installation (aside from lead-acid based microgrids). This was true not just in the Pacific but to a lesser extent internationally. This "maturing" industry status significantly influences a number of the issues raised in this assessment.

For example, many of the defects and control and earthing commissioning issues observed in BESS on Ha'apai (OIREP), and the first load-shifting BESS on Rarotonga (CIRESP), were due to low product maturity. Even the selection of the first BESS on Rarotonga as load-shifting, when a grid stability BESS would have offered more initial benefit, was driven by the technology maturity status at the time (BESS inverters did not have proven, in-built grid support functionality, and the utility control system was in the early concept stage only).[10]

Currently, the situation is vastly different. There is a wide range of fully integrated, type-tested products, satisfying dedicated BESS standards for performance, safety, and environment. There are many large-scale BESS specific inverters with integrated control capability that provide all grid support functions. There are microgrid controllers supporting battery management and distributed energy dispatch. There are a large number of contractors experienced with selection, design, and installation of BESS systems at various scales.

The transition in the BESS industry since 2015 is illustrated in Figure 5.1.

There is still significant progress to be made, and particular gaps remain in product standardization, end-of-life treatment, clarity on emergency services response requirements, consistency in definitions of control capabilities, and diesel-off operational capability. As such, procurement of BESS still requires a high degree of specification and prescription and stringent quality control requirements. Importantly, the procurement specifications to date have placed a strong emphasis on the quality and rigour of the factory acceptance test (FAT) to resolve technical issues before equipment is brought to the site (and issues become much harder to solve). While this has been effective, strong enforcement is necessary. There are still issues arising at site (typically less significant, including inadequate corrosion protection of minor parts or tuning of control system parameters). There is also potential to expand requirements for more standard equipment such as switchgear.

For future projects, it is most critical to understand the evolution and remaining gaps. The most important of those gaps is product standardization. Understanding that BESS offerings still vary widely and may not necessarily include all features required for a project (or may include more features than required) ensures specifiers give detailed consideration to the required functionality and operating environment and specify or select applicable standards or requirements accordingly. For example, whether active fire suppression is specified for a BESS unit

---

[10]   In addition, maturation has significantly reduced the cost of BESS, both directly (capital cost), and indirectly: in both OIREP and in the Cook Islands outer islands, the adopted lead-acid batteries have a shorter life (~8 years) compared to lithium batteries now available (~15 years). Lithium batteries also require less than 50% of the space.

Figure 5.1: Maturation of BESS Industry

	Project Milestone	Technology capability	Standard	Price
2015	OIREP and CIRESP Outer Islands Contracts awarded	Small, off-the shelf units capable of grid forming for micro-grids	Lead, acid battery standard and various routine electrical/civil/structural standards or borrow from PV substation or vechicle standards	Lead, acid batteries significantly cheaper than Lithium
2016		Custom integrations of battery modules, containers, inverters and balance of plants Minimal grip support functions. External bespoke controls		Lithium price competitive for constrained sites
2017	CIRESP GEF BESS Contract awarded (1MW/4MWh)	Improved quality control, type testing by some suppliers		Energy price continued to drop at about 7% annually
2018	CIRESP GCF load shifting BESS Contract (2MW/8MWh)	Integrated BESS inverter controller capability improving for VF control	Substantially improved fire safety test standards	
2019	TREP Contracts awarded (>16MW/40MWh) CIRESP GEF and GCF Load shifting BESS completed		Development of integrated BESS standards, definitions, tests and specifications	
2020	CIRESP GCF Grid Stability BESS Contract awarded (6MW/3MWh)	Widely available modular components including fully integrated cabinets/kiosks/containers	Environmental labelling standards Dedicated lithium-based BESS standards	
2021		More widespread compliance of BESS with control functionality and safety standards Increasing availability of dispatch control for hybrid systems		COVID-19 logistics, raw material costs, fire concerns, pause price curve

BESS = battery energy storage system, CIRESP = Cook Islands Renewable Energy Sector Project, GCF = Green Climate Fund, OIREP = (Tonga) Outer Island Renewable Energy Project.

Source: Author's experience in delivering BESS and microgrid projects over this period

or just passive prevention of fire propagation is a decision that is likely to be affected by the location, proximity to other infrastructure, the capability of emergency response, public perception, and budget.

Importantly though, it is now apparent that BESS can offer a full suite of grid support functions allowing stable operation of small, medium, and large isolated networks with high renewable contribution, even without synchronous (diesel) generation online (subject to some limitations as discussed in the next section).

5.1.2 Technology: Improving Alignment Between Technical Considerations and Decision Making

Technical assistance in the preparation of feasibility, specification, tender evaluation, and construction supervision has been a core component of all case studies. Stakeholders observed that experienced and multi-disciplinary technical assistance represented a significant success factor, particularly where there was continuity through the project life cycle.

However, it was also observed that there is still significant gaps in the knowledge of various stakeholders about the technical issues associated, particularly with medium to high renewable energy hybrid systems, BESS, technology selection, and control systems requirements. Similarly, technical assistance did not always properly understand the drivers and requirements of local stakeholders. This inherently impacted decision-making and resulted in some outcomes that did not necessarily match expectations. Examples follow:

(i)   For CIRESP small grids on outer islands, stakeholders have expressed the following concerns about some islands not meeting renewable energy contribution targets and selected sealed lead acid battery technology being outdated, not providing expected autonomy, and not likely to meet expected lifetime. Such concerns, while legitimate, do not reflect the initial decision-making process (Entura 2015):

   (a)   a government recommendation for commonality of equipment with existing northern group projects (for ease of maintenance);

   (b)   fire safety risk of immature lithium-ion battery projects on remote islands;

   (c)   significantly higher capital cost of lithium-ion batteries at the time of procurement (2015);

   (d)   uncertainty in expected performance of the plant subject to solar resource and other assumptions (three of the four projects are actually performing within the uncertainty bounds of the original projections; and

   (e)   sensitivities for load growth and the likely impact on renewable energy contribution.

(ii)   For CIRESP, as described in section 3.4, the case for the first load-shifting BESS was used as the basis for three more units of BESS, one of which was later changed to a grid stability BESS to enable the required additional solar PV generation (which would not have been enabled by the selected load-shifting BESS).

(iii) As an example from TREP, section 3.3 describes how Vava'u project design sizing for the solar PV did not correspond to the efficient utilization of the BESS. Project performance could have been improved through further economic optimization using energy balance modelling. Additionally, the project only resulted in a modest 7% increase in renewable energy, which is relatively small compared to the progress needed toward renewable energy targets. In hindsight, a significantly larger solar PV array may have been preferable (and is currently under consideration as a change in scope).

(iv)   Also, in TREP, section 3.3 describes the change from decentralized to centralized BESS for Tongatapu to maximize efficiency and achieve grid stability.

A related issue is that there was relatively low community literacy on the impacts and opportunities that come with a renewable energy transition. This is not unique to Pacific SIDS and is the case in most markets. However, it is important that government and utility stakeholders understand how this can impact the success of its programs, particularly in relation to perceptions about customer tariffs and opportunities to participate in energy generation (e.g., rooftop solar PV). Customer perceptions about tariffs are a recurring theme in the case studies and other projects and may benefit from coordinated and fact-based responses.

For example, in Rarotonga, the BESS projects required the installation of an additional 8 MW of solar PV to obtain the expected benefits. Initial public consultation by the utility indicated a high level of customer interest in embedded generation (Te Aponga Uira, 2019) (small- to medium- scale rooftop solar PV) based on a feed-in-tariff or PPA, or offsetting consumption tariffs (net metering). Initial planning, therefore, progressed based on a customer embedded generation model. However, consumer sentiment did not consider the complex control requirements, the infeasibility of continuing with high historic generator tariff arrangements, or the impact of high land lease costs. While planning has now adapted, this has contributed to delays in implementing solar PV on Rarotonga.

The project technical advisors have fielded a number of other frequently asked questions from stakeholders during project feasibility and delivery. In addition to the themes above, these commonly relate to optimal sizing of BESS, benefits and requirements of being able to operate without synchronous diesel generation operating (i.e., grid-forming BESS), and how to avoid constraining solar PV generation (and the associated losses).

Based on the above, information required to address key knowledge gaps includes:

(i)   Are distributed or centralized BESS better?

(ii)   Is diesel-off operation required?

(iii)   How do you size a BESS?

(iv)   Is a BESS better employed for load-shifting or grid support?

(v)   Is curtailment of renewable energy a problem?

(vi)   Control system requirements for different stages of renewable energy transition

(vii)   BESS control functional requirements

(viii) Cost–benefit of dominant RE generation technology (solar vs wind)

(ix)   What opportunities are there for customer investment and/or participation?

(x)   Should customer tariffs be reduced?

There is a range of other issues that arise. However, these are considered key factors in informing early-stage decision-making and aligning expectations of project performance and risk. To this end, a series of brief infographics have been included to support increased understanding and provide an improved basis for rapid and/or early-stage decision-making. These are not intended to replace detailed analysis and technical assistance. They are displayed in the following figures.

Information to support decision-making, such as given above, is insufficient without a structured model for informed decision-making such as the responsible-accountable, consult, inform (RACI) model (Costello 2012). Current decision-making processes, at the project initiation stage, including concept development, tend to be more ad hoc, consistent with that observed in other developing countries (Hirmer, et al. 2021). For hybrid energy projects in isolated grids, stakeholder engagement during initial project selection and definition may benefit from utilizing the following model:

(i)   **Sponsor:** Allocate based on country structure (likely utility, government energy department)

(ii)   **Government – energy ministry:** Responsible and accountable—a strategic decision-maker

(iii)   **Government – financial ministry**: Consult

(iv)   **Utility:** Consult

(v)   **Regulator:** Inform

(vi)   **Customers:** Consult and inform

(vii)   **Landowners:** Consult and inform

(viii) **Regional developers, contractors and investors:** Inform

(ix)   **PMU, TA:** Consult

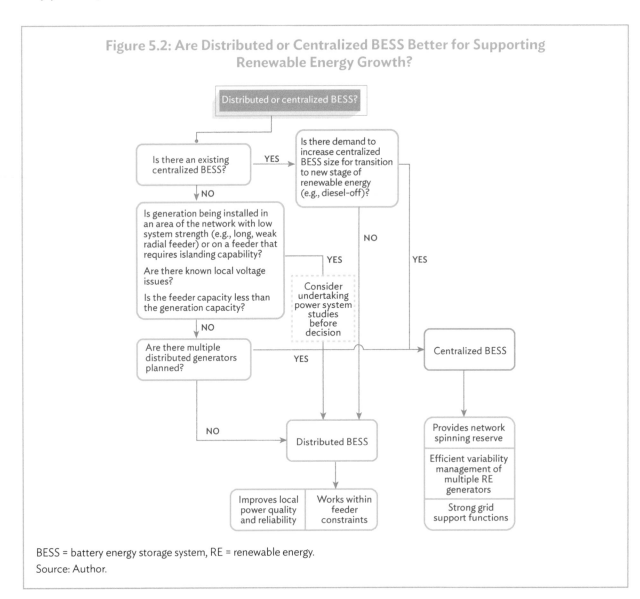

Figure 5.2: Are Distributed or Centralized BESS Better for Supporting Renewable Energy Growth?

BESS = battery energy storage system, RE = renewable energy.
Source: Author.

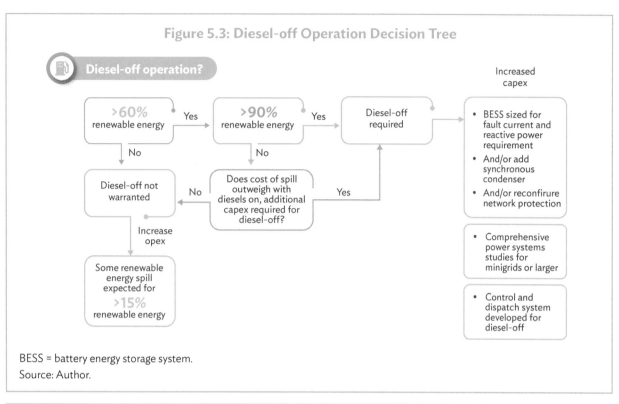

Figure 5.3: Diesel-off Operation Decision Tree

BESS = battery energy storage system.
Source: Author.

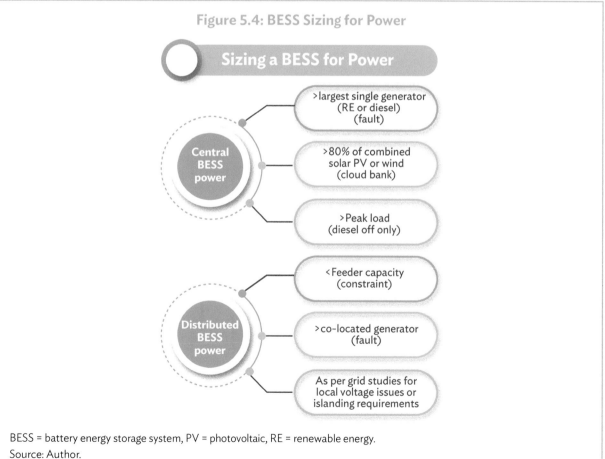

Figure 5.4: BESS Sizing for Power

BESS = battery energy storage system, PV = photovoltaic, RE = renewable energy.
Source: Author.

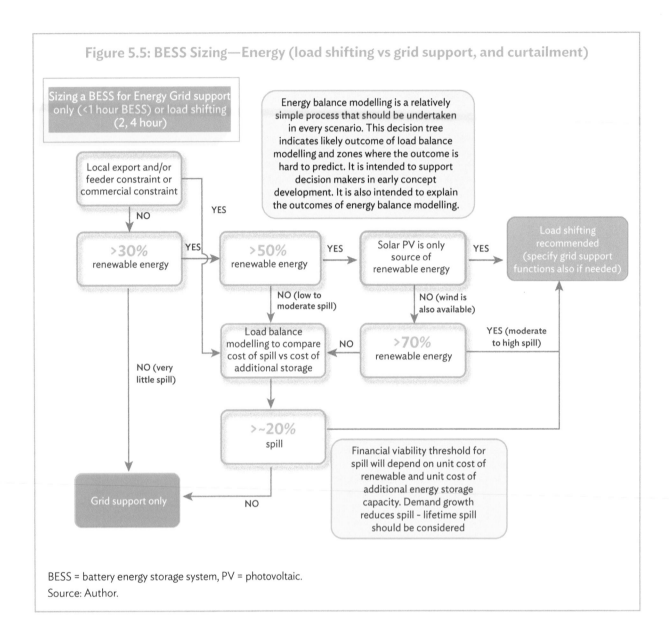

Figure 5.5: BESS Sizing—Energy (load shifting vs grid support, and curtailment)

BESS = battery energy storage system, PV = photovoltaic.
Source: Author.

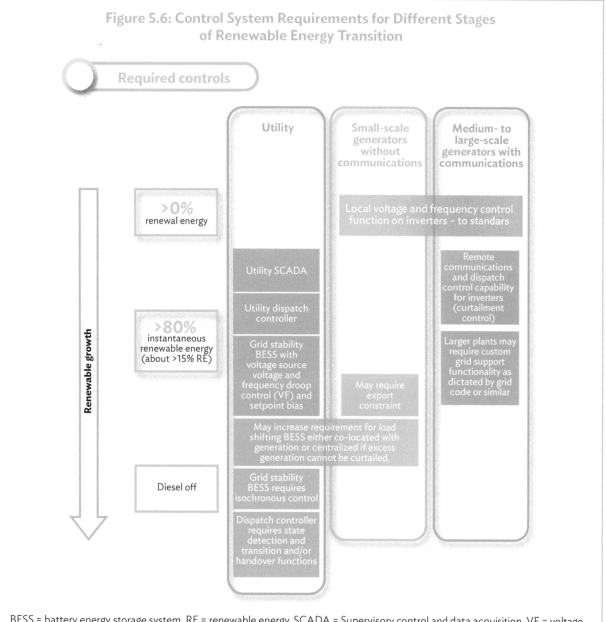

Figure 5.6: Control System Requirements for Different Stages
of Renewable Energy Transition

BESS = battery energy storage system, RE = renewable energy, SCADA = Supervisory control and data acquisition, VF = voltage frequency mode.

Source: Author.

Figure 5.7: BESS Control Functional Requirements

BESS
INVERTER CONTROL MODES

Grid following mode (PQ)
- Control loop following externally provided active power and reactive power set points
- Voltage and frequency set by grid
- Reactive power set as power factor or voltage
- Typically active power controlled for load-shifting
- Slow set-point update (1–2 second interval)
- Limited role in grid support
- Not able to form a grid (requires synchronous generation or grid forming BESS)

Basic load shifting applications

Can do grid support with external controller but limited due to update rate

Grid Forming (VF)
- Control loop acts on error in grid voltage and frequency
- Fast response <50 milliseconds

Grid forming droop
- Open loop control (active and reactive power proportional to error in grid frequency and voltage)
- Active power and reactive power (or power factor) can be controlled to external set-point
- Reactive power set as power factor or voltage
- Typically active power controlled for load-shifting

Basic load shifting applications

Can do grid support with external controller but limited due to update rate

Grid forming isochronous
- Closed loop control (active and reactive power controlled to bring grid frequency and voltage error to zero)
- Active power and reactive power (or power factor) cannot be externally set (controller will just counter to maintain frequency and voltage)
- Should only be one isochronous source on the network to avoid instability

Basic load shifting applications

Can do grid support with external controller but limited due to update rate

BESS = battery energy storage system, PQ = active power-reactive power mode, VF = voltage frequency mode.
Source: Author.

Figure 5.8: Cost–Benefit of Dominant Renewable Energy Generation Technology (solar vs wind)

Lower cost of energy

Higher storage requirement

Nighttime generation

Higher cost of energy

>50% renewable energy adds substantial storage requirement and increasing cost

Solar typically dominates

Can delay storage cost >50% renewable energy

Source: Author.

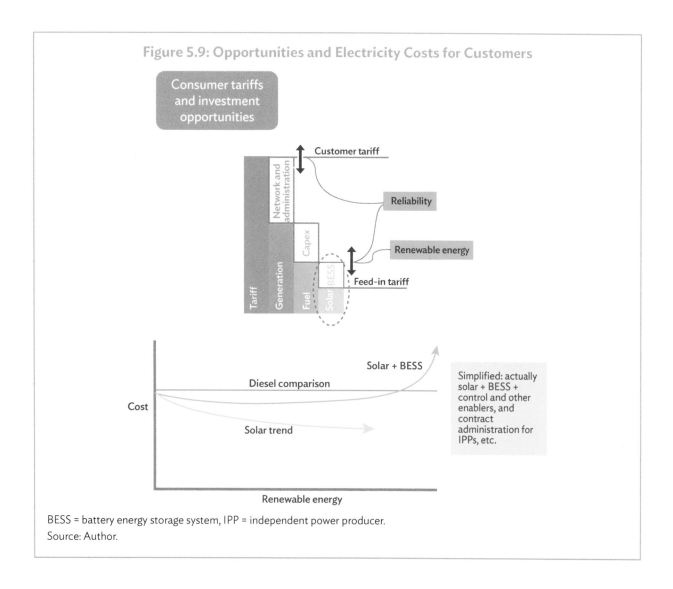

Figure 5.9: Opportunities and Electricity Costs for Customers

BESS = battery energy storage system, IPP = independent power producer.
Source: Author.

5.2 Standards, Safety, and Environmental Challenges

Standards, safety and environmental challenges were not frequent topics in most of the subject matter expert interview responses. The exception was end-of-life equipment treatment that was raised as a concern considering Pacific SIDS limited capacity for waste disposal or recycling. This also included the suitability of local codes for dealing with new products, including shipment and transport of what may be classified as hazardous waste.

Nevertheless, these areas were of concern for the technical advisors, reflecting broader concerns in other countries and markets. The issue of standards for BESS is covered well in the existing literature (section 5.3), and will not be expanded on here except to highlight as per section 5.1.1 that there is a need to maintain a watch on the latest standards development and incorporate into project specifications.

Standards are increasingly available to deal specifically with safety challenges. In particular, electrical and mechanical safety issues are well covered. Significant progress has been made with respect to fire and explosion risk also. Despite BESS fire events still occurring (Australian Broadcasting Corporation 2021), the frequency is decreasing. The combination of passive safety design measures (such as firewalls limiting propagation, venting upward) and emergency procedures (primarily preventing access and waiting) means the consequences are also less severe.

However, fire safety should still be a high priority consideration for Pacific SIDS. Aside from the potential loss of infrastructure, there is a risk of the conflagration to other equipment at many sites with space constraints and limited local capabilities for emergency response and health care. Provided that specifications and construction supervision ensure the stringent application of current standards, the key gap for Pacific SIDS is the emergency response. The key finding from recent fire events (McKinnon 2020) applicable here is the requirement for appropriately trained first responders. Lithium BESS fires have unusual behavior in which they can cycle (alternatively heating and appearing dormant) that can persist for many hours or even days. Injuries have occurred where first responders attempt to access BESS when the fire appears out. All BESS specifications and delivery should therefore include emergency services within the training delivery, using an accredited training program.

Environmental challenges (specifically end-of-life treatment) are so far less well covered within standards. Labelling standards (International Electrotechnical Commission, 2019) should be followed to facilitate end-of-life materials separation and handling. However, otherwise, limited reliance on standards is possible. Product design life (10–15 years) also means that for EPC-type projects, reliance on the contractor to include end-of-life treatment is not practical. An Original Equipment Manufacturer (OEM) recycling programs are sometimes offered. However, they are not widespread and typically rely on the customer for shipment back to the recycling facility. This issue is not unique to Pacific SIDS (or batteries; similar issues exist for solar PV modules), and rapid industrial progress is expected to develop regional recycling facilities due to increasing demand. However, there remains a gap in getting batteries off the island at the end of life and to such facilities. There is an opportunity for funders to guarantee funding for such activities as means to ensure desirable environmental outcomes and support the uptake of this technology.

This may be coupled with a review of local regulations to support the appropriate management of batteries at the end of life. Finally, because there is now such a wide selection of valuable international standards, there is an associated burden for small utilities or SIDS to purchase, understand, and maintain this resource. This is not particular to BESS or microgrids that may reduce over time as this equipment becomes commonplace. However, in the meantime there is expected to be a reliance on technical advice to leverage standards.

5.3 Procurement Challenges

5.3.1 Procurement: Constrained Procurement Options

Except for some smaller items of work, all procurement packages adopted the International Competitive Bidding Procurement Method under ADB's 2015 Procurement Guidelines utilizing the ADB standard bidding documents for procurement of Plant (see Figure 5.10). The following concerns were raised following the completion of procurement activities on the case study projects:

  (i)   Project managers, utility representatives, and contractors all identified concerns about low flexibility to adjust the specification or offer to optimize outcomes. This concern was generalized across all

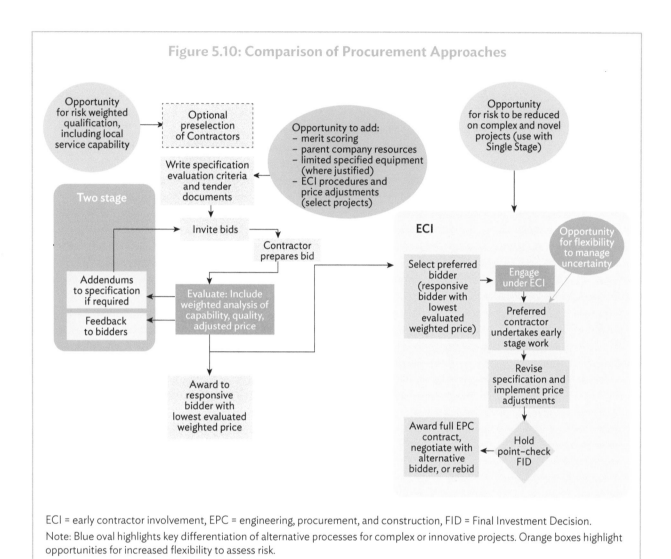

Figure 5.10: Comparison of Procurement Approaches

ECI = early contractor involvement, EPC = engineering, procurement, and construction, FID = Final Investment Decision.

Note: Blue oval highlights key differentiation of alternative processes for complex or innovative projects. Orange boxes highlight opportunities for increased flexibility to assess risk.

Source: Author.

case studies and reflected discussions by stakeholders throughout the delivery of projects over the implementation period.

(a)   Example: A bid may be evaluated as having the lowest price and a strong technical offer, but with one or two isolated deviations that could not be considered minor. Such deviations may be commercial: For example, in TREP outer islands tender, a bidder requiring a limitation on the period during which claims can be brought against the contractor was rejected; or technical: For example, in CIRESP one bidder offered a compliant solar–BESS–diesel package that did not allow the new diesel generator to integrate with existing diesel generator controller (it could only run with the BESS controller online, which may not always be available), and was rejected. Such bids must be rejected, even though the matter may be easily resolved through negotiation and perhaps minor price adjustment prior to award. In some instances, such issues can be resolved through a two-stage procurement procedure (see Figure 5.10), which allows for feedback to the bidder and adjustment of their offer after the first stage. A two-stage process adds significantly to the timeline and effort for procurement and is normally reserved for technically complex procurement activities. However, this can be somewhat offset by savings in the final bid analysis.

(b) Example: For the implementation of a network-wide control system that incorporates various types of legacy and new or future distributed and centralized generators and various functions and commercial levers, a full specification prior to tender may not be practical, and often, complete drawings and data sheets of existing equipment are not available. With this concern in mind, TAU on Rarotonga, Cook Islands, opted to independently specify and procure the network control system (independent of the grant-funded project). Some progress was made. However, this independent process was not concluded due to changes in the priorities and approach of TAU and the complexity of the problem. Regardless, it highlights the limitations of the standard procurement options.

The approach of TAU may have been better supported within the CIRESP, had a more flexible procurement option been available. In particular, an early contractor involvement (ECI) (State Government of Victoria, 2021) approach is likely to suit this situation. ECI involves the conduct of tenders based on an initial specification to select a preferred contractor. Prior to engaging the preferred contractor for the full EPC project, the preferred contractor is initially engaged to complete the preliminary design and specification. This allows preferred contractor to collect more detailed information and the employer to collaborate in the design decision process, better account for the financial implications of alternative solutions, and review the financial investment decision. Subject to well defined price adjustment mechanisms to account for any differences between the final design and the tender, the employer may then award the full EPC contract to the preferred contractor (or if necessary go back to market or not proceed with the work). This applies similarly to complex and innovative projects, or where the technology is relatively new and is best suited to collaborative development between the employer and contractor. It is also applicable for BESS, where preselection of inverters and detailed grid stability studies using selected inverter models may be warranted prior to procurement for the full EPC package (incorporating the selected inverter).

These examples would not be addressed through a two-stage procurement process. Two-stage procurement allows for a preliminary review of the bidder's technical solutions in Stage 1, with feedback to bidders to be addressed in Stage 2. This is closer to an ECI model but does not provide the level of depth needed for site investigations, data sharing and detailed control modelling, or techno-economic optimization and decision-making. Typically bidders will not contemplate this degree of effort during an open bidding process, nor would it be practical for the employer to manage such detail with multiple bidders.

(ii) On both the Cook Islands and Tonga, stakeholders expressed concern about the inability to utilize weighted or quantifiable evaluation criteria or otherwise consider factors that may not be predictable when preparing bidding documents. The "lowest priced substantively conforming bid" basis for selection was considered overly simplistic. Again, this was generalized across all case studies but was focused on two projects, one a large BESS procurement on Rarotonga under CIRESP, and the other a mini-grid project under OIREP, where this issue was perceived to result in contract award to higher risk bidders and where these risks were borne out in implementation.

(a) Example: Setting a relatively low minimum experience qualification criteria is necessary in a niche market with relatively few experienced contractors to increase the chance of an outcome from the tender. However, it may be found during evaluation that two bidders, both meeting the minimum threshold, have significantly different experiences, which would strongly affect the project risk profile. For example, under one of the case projects, early BESS tenders recognized the limited industry experience and set a threshold of one (1) BESS project within the last five years with a similar capacity. Bidder A qualified with a single BESS project experience, while Bidder-B had more than five such projects. The standard evaluation procedure does not readily allow this to be considered.

(b) Example: A similar and common scenario is two technically equivalent bids, but one contractor is based regionally (similar time zone and single flight to country), and the other contractor is based on the other side of the world. This situation has arisen in every one of the case studies. Again, the project risk is significantly reduced by the regional contractor (this has been clearly demonstrated with COVID-19), but this is not considered in the evaluation. Also, while the regionally based contractor may have some travel cost advantage, this is a small factor in the total bid price and may be offset by smaller market and higher basic operating costs, such that local presence will not typically ensure the lowest bid price. As such, the lowest bid price cannot necessarily act as a surrogate for this risk factor.

(c) Example: In one of the mini-grid projects, two bidders may offer equipment meeting all specifications. Bidder A included a BESS that achieved the specified usable energy storage, while Bidder B's energy storage was about twice that specified. Bidder B's selection was driven by other factors such as their supplier's standard equipment sizing and capability to meet short term power requirements. Both bidders were equally technically qualified but Bidder B provided significantly more qualitative flexibility, robustness against degradation, and potential for future expansion. Similarly, on grid stability BESS, one bidder offered three transformers compared to other bidder's single transformer offer. Both were compliant. However, the three transformer case offered better redundancy when considered in conjunction with the specific modularity of the BESS. While in this case, it would not have been a determining factor due to the price difference of the BESS (irrespective of the transformer solution), this type of risk could differentiate bidders. Still, it would not be considered in deciding the award.

The inclusion of merit-point evaluation criteria in procurement design, which can consider relative risk and opportunity of bids for qualifications and technical offer, would offer an efficient way to address these concerns.

(iii) Stakeholders, particularly on the Cook Islands, also expressed concern about the constraints for selecting a specific equipment brand and/or type, particularly where this would benefit O&M or training requirements for utilities managing multiple equipment types. On Rarotonga, Cook Islands, there are now three BESSs, each using a different battery module and different inverter technology.

(iv) In both CIRESP and TREP, stakeholders questioned the limitations imposed by requiring only capability and experience of the bidder, and not its parent, direct subsidiaries or affiliates be considered for evaluation. This was counter to the typical industry model of establishing special purpose companies to deliver projects, where either experience or finance are separated into different legal entities even though they are delivered under a single umbrella company with the same resources.

Despite the issues raised, stakeholders acknowledged strong support from the ADB procurement and project administration, throughout the projects. The authors note that the flexibility sought in the above comments requires careful design and coordination during procurement preparation. However, it is clear that these issues are important to stakeholders. In some cases, elaborate systems of adjustments for battery and inverter integration and warranty duration were written into the evaluation criteria to try to satisfy stakeholder concerns. However, explicitly addressing the underlying concerns by using a merit-based assessment in the evaluation is the preferred approach.

As noted previously, the case study projects were subject to ADB's 2015 Procurement Guidelines. In 2017, introduced a new Procurement Policy[11] and associated Procurement Regulations[12] applicable to all projects commencing after July 2017. The new policy strengthened ADB's commitment to achieving fit-for-purpose procurement solutions and increased the focus on designing project and context specific procurement solutions through a new strategic procurement planning (SPP) exercise.

Conducted in a comprehensive manner at the beginning of a project, the SPP process can be used to research and identify the most suitable procurement approaches to engaging with the private sector to purchase goods, works, nonconsulting and consulting services. Subject to adhering to ADB's core procurement principles,[13] it can be used to support the incorporation of bespoke methods of engagement, such as ECI, together with industry-specific contracting forms and innovative evaluation approaches with the goal of developing a fit-for-purpose procurement strategy for a project.

SPP involves a detailed analysis of the needs of the project owners, the capacity of the market to meet those needs, the expectations of project stakeholders and an analysis of internal and external risks to successful delivery. Using this information, various procurement options are developed and compared with the optimal solution selected and subsequent work to realize the project outcomes undertaken.

In response to the evolving nature of procurement globally, the 2017 Policy also adopted several improvements that address directly the issues observed in the case study report, as well as other common problems. These include (i) the option to include merit-point evaluation criteria as part of the evaluation for all procurement packages; (ii) the ability of bidders to rely on the experience and financial resources of their parent companies; (iii) an increased focus on contract management; and (iv) an enhanced focus on environmental, health and safety requirements. All these aspects can be built into the analysis conducted during the SPP process.

Adapting the ADB standard bidding documents to address the concerns and introduce recommendations discussed here does require strong and early engagement with ADB, and a clear intent to develop bidding documents that address specific stakeholder concerns. A well informed and experienced project management, procurement and technical team are highly beneficial in facilitating this process in a timely manner and ideally such development can proceed in parallel with early-stage feasibility so as not to significantly impact the project schedule.

5.3.2 Procurement: Aligning Interrelated Contracts

A scenario took place on both CIRESP and TREP main island projects that are expected to commonly arise for this scale of project. The BESS component was designed to support the installation of more solar PV. Delivery of the BESS and solar needed to align because they could not operate effectively until the BESS was completed. The BESS would sit idle (depreciating) until the solar was completed, or alternatively, the solar would be producing energy that would get spilt (costing the IPP); or cost the Utility for the power they could not use. To further complicate this, both solar and BESS needed some grid upgrades or control system (SCADA) changes that were to be delivered by the utility.

In TREP, on Tongatapu, the solar and BESS were delivered concurrently to attempt alignment, with the completion of the BESS planned for 3-6 months ahead of solar. However, various factors, particularly COVID-19 delays, meant that both projects were delayed. The BESS experienced greater delays, impacting the commercial operation date of the privately financed 6 MW solar IPP project, exposing the utility to potential losses on take- or-pay[14] PPA contracts with the IPP for energy it could not use.

---

[11] ADB 2017. ADB Procurement Policy. Goods, Works, Nonconsulting and Consulting Services. Manila
[12] ADB 2017. Procurement Regulations for ADB Borrowers. Goods, Works, Nonconsulting and Consulting Services. Manila
[13] ADB core procurement principle are economy, efficiency, fairness, transparency, quality and value for money.
[14] Take or pay contracts are largely unavoidable in isolated grids with physical purchase power agreements.

In CIRESP, the BESS was planned to be fully delivered before contracting new solar (effectively introducing 12–24 months lead time for the BESS). However, the first shifting BESS was delayed by 18 months while the contractor resolved precommissioning issues, and the grid stability BESS was delayed approximately 2 years) because of retendering and COVID-19. However, despite these delays, only about 1 MW of the required additional solar generation has been completed, with the remainder not yet commenced. As mentioned in section 3.4, the delays in the solar were partially due to reluctance to commence without having the grid stability BESS in place. Consequently, some of the load-shifting BESS capacity may sit idle for at least 2 years before use.

Reflection on this issue has identified that using such models, misalignment is high risk and should be factored in as a likely outcome. Indicatively, misalignment may occur as follows:

(i)   Up to +/-3 to 6 months difference: likely (typical matters such as minor delays in one or the other contract)

(ii)   Up to +/-9 to 12 months difference: possible (event such as protracted negotiations, change in site, unforeseen conditions etc.)

(iii)   Up to +/-18 months difference: unlikely (event such as mis-procurement, contractor default, disaster, pandemic, etc.)

For the TREP case, the allowed lead time of 3-6 months would have usually sufficed as a contingency, and it was only the rare occurrence of COVID-19 that prevented this, with the contingency ensuring less severe impacts. For CIRESP, the outcome of the approach is less ideal due to the delay in realizing project benefits from additional solar PV.

Mitigation measures should be planned accordingly and include both reasonable contingency lead time for enabling technology and commercial or technical measures to mitigate the impact of misalignment. These should be delivered in conjunction with tight management of all preparation and contracts. Suggested mitigations are:

(i)   Plan to deliver BESS, solar and grid concurrently, but with 3-9 months lead time on the completion date of the BESS as contingency (weight toward early delivery of BESS and grid as impacts are less severe).

    (a)   For centralized grid stability BESS, a plan for utilization for spinning reserve will reduce diesel generator O&M costs even before solar PV generation is commissioned.

    (b)   If both BESS and solar are delivered in accordance with this plan (BESS 3-9 months early), return on investment may be less than if they had commenced operation together. This can be factored into the feasibility studies.

(ii)   Solar PV contracts may have a hold point included prior to procurement of major items, pending commencement of BESS or grid contracts. Thus, if award of BESS and/or grid contracts is delayed such as by a misprocurement, solar PV costs can be minimized)

(iii)   If the BESS or grid are completed <3 months after solar PV, solar PV can export up to hold point 1 (determined by grid studies, maybe 10% output).

(iv)   If the BESS or grid are completed <9 months after solar PV, manual operation of solar PV curtailment limit may be warranted to better utilize energy and/or save fuel. Development of operating procedures will be required, and some reduction in grid reliability may be considered to increase utilization.

(v)   If the BESS or grid are completed <18 months after solar PV, further intervention may be required, particularly if this is known early, there may be opportunity to implement further technical or commercial mitigations.

(vi)   Use a single management or decision-making team to coordinate projects, even if delivered under different funding initiatives.

Stakeholders also raised, as possible mitigation, ensuring that interdependent works are delivered through a single funding source or delivery model (e.g., solar and BESS delivered together as a single IPP project). This has merit as it pushes interface risk to the contractor and is typical of small- and medium-scale projects that did not have the above issues. However, when adding grid connection / upgrade and control systems elements, it may not be practical or efficient to include them all under a single package. This option should be considered on a case-by-case basis.

5.3.3 Procurement: Managing Contractor Incentives

In some projects, underperforming contractors were identified as a key challenge. In particular, this was driven by a lack of allocation of contractor resources to complete projects in a timely and high-quality fashion (this appeared to be more common where the contractor had underpriced the work). Project managers sought increased power to incentivize Contractors, particularly during execution.

All the projects used ADB standard bidding documents and generally followed guidelines for advance payment and other payment milestones, guarantees and securities, and liquidated damages. In consideration of the nature and value of some of the issues arising in BESS projects, these were not always sufficient.

Liquidated damages were challenging to enforce, requiring very high levels of diligence in project management, record keeping and legal support, and strong engagement with the contractor in order to maintain a working relationship. For example, one of the first BESS experienced approximately 18 months in delays attributable to the contractor. The liquidated damages were insufficient to incentivize the allocation of additional resources to the project. When delays reached 12 months, the delay liquidated damages cap was reached, and there was no further incentive available to motivate the contractor. Works were finally completed approximately 18 months late.

Employer step-in rights to take over complete works or rectify defects were considered in this case. However, these were deemed likely to increase delays due to the highly specialized nature of the equipment and specialized skill sets needed to take over work from the contractor.

In general, these concerns were seen as reducing with technology maturity (see section 5.1.1), and increased availability of experienced contractors and modular products mean the above case is less likely to reoccur. However, minor adjustments in contract terms may be warranted. These should achieve a balance between the significant residual risk carried by battery and control projects up to commissioning (integration and performance issues can arise late in implementation and be difficult or expensive to resolve) and the contractor's costs that typically occur early in the project for procurement of major items.

The following settings, which represent minor adjustments compared to ADB standard bidding documents may be warranted. These are reflected in Figure 5.11.

(i)   Subject to a risk assessment and advance payment guarantee, an advance payment of up to 20% should be considered (compared to the typical 10%). This better reflects a contractor's early procurement costs, moderates contractor's financial resource requirements, and provides more scope for subsequent contingencies discussed below.

(ii)   Completion payment and commissioning payment milestones are recommended to be 10% and 5% respectively. This is a significant residual amount for Contractors (who would have expended approximately 100% of costs by completion). However, given the propensity for issues at this stage, this can be substantiated. Break-up as follows:

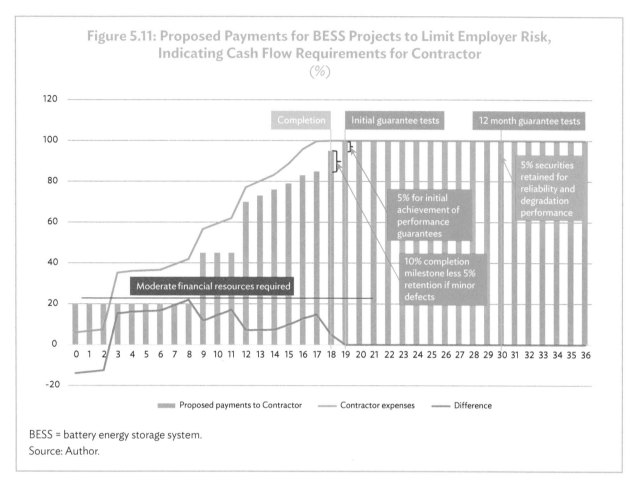

Figure 5.11: Proposed Payments for BESS Projects to Limit Employer Risk, Indicating Cash Flow Requirements for Contractor
(%)

BESS = battery energy storage system.
Source: Author.

(a)   At completion date, 10% milestone paid . It is also recommended to retain an additional amount if minor outstanding defects are at completion. This should be sufficient to cover rectification of such defects by others if the contractor does not subsequently carry out the work. Alternatively, performance security could be adjusted by a similar amount.

(b)   Within 30 days of completion, initial Functional Guarantee tests completed and 5% milestone paid (less any performance LDs from initial tests).

(c)   At 12 months, second Functional Guarantee tests were conducted (checking performance degradation of BESS and plant availability). However, holding a payment milestone for this long is a significant barrier for many Contractors, therefore, the use of securities to manage this risk is recommended. Any performance LDs payable during commissioning may be subject to a re-test after 24 months, allowing Contractors to implement and demonstrate improvements, and reduce the total performance LDs payable.

(iii)   In terms of securities:

(a)   Standard advance payment bank guarantee, reducing over contract period proportional to subsequent payments.

(b)   Performance guarantee of 15%, reducing to 10% on completion and reducing further to 5% following successful completion of functional guarantees at 12 months after completion. Five percent residual should be retained (not reducing) for 2 years post completion. This would normally need to coincide with a 2 year defects liability period.

The above measures are likely to come with some cost, which is not negligible. The above guarantees may add 0.5% to the contract price.

In addition to the above, OEM warranties should be provided and automatically assigned to the employer (without needing the cooperation of contractor) at the end of DLP or at the request of the employer.

Notably, the above measures (along with existing protections in the contract) are ineffective without strong project management, including diligent record keeping, timely inspections and issuance of notices, management of guarantees, securities and warranties, post completion monitoring, and willingness to enforce protections.

5.4 COVID-19 Challenges

5.4.1 COVID-19: Insurance Costs and Availability

A steep rise in insurance costs for Pacific renewable energy projects has occurred since 2019. This impacted the case study projects because COVID-19 delays required an extension of all-risks type insurance cover. However, COVID-19 does not necessarily explain the price increases, and other underlying factors may be present.

Based on follow-up discussions with stakeholders, insurance cost increases during implementation have been anywhere from 300% to 3000%, and there are instances where actually obtaining insurance has been in

doubt. To date, insurance has been able to be maintained at rates up to about 3% of capital cost. This has had a considerable impact on project return on investment.

Stakeholders attributed cost increases to any of the following factors:

(i)   COVID-19-related logistics and shipping issues;

(ii)   increased cyclone impacts on Pacific islands in recent years;

(iii)   reduced competition in the niche insurance market representing the Pacific; and

(iv)   concerns about perceived battery fire and explosion risk arising from incidents in other markets.

While each of these is a plausible and real issue, this analysis has not been able to determine if they or another factor has actually caused the price rises. However, several of these factors may persist beyond COVID-19 and may also affect operations periods. These represent a significant risk to future projects viability.

Options to mitigate the risk in commercial markets have been extensively explored by a range of stakeholders without success. Therefore, other mitigations may be required. Stakeholders raised the following options:

(i)   Considering their portfolio of investment and interest, an ADB or other funding agency backed insurance scheme may be one option. This would not only offer potential savings but reduce administrative time and cost in sourcing and negotiating insurance on a project by project basis. However, this approach may be considered contrary to market competition for insurance services or may not be a good long-term fit for the operations period of projects delivered on a short-term basis.

(ii)   Technical change to the project structure, such as climate risk assessment, inland and elevated or protected siting options, or more stringent technical specifications. However, these measures are already undertaken and essentially act as prerequisites for access to insurance (though continuous improvement is necessary as standards or practices change). There are limited options to further tighten technical specifications against insurable events.

This issue remains an outstanding risk for current and future projects.

5.4.2 COVID-19: Remote commissioning

Many stakeholders identified modular, fully integrated BESS equipment as highly desirable, avoiding bespoke, customized installations and the various issues this creates. However, stakeholders also identified that such products are typically subject to constraints on who can work on them. Several case study projects are currently delayed because the authorized staff are based in Europe and cannot attend the site due to COVID-19 restrictions.

Even without COVID-19, fault rectification represents a barrier, with respondents indicating a need to liaise with authorized representatives overnight (due to time difference) to resolve minor problems and any more substantial problems again requiring travel from the equipment supplier or agent to the site.

Both TAU and TPL have considered staff undertaking authorized training. However, this is not available for all products and is not practical for limited resources to cover requirements for multiple equipment suppliers.

While several BESS manufacturers have taken steps toward allowing remote commissioning (since this affects not just Pacific projects during COVID-19), progress has been slow, and it is not yet clear if this will be a viable option in the future if all warranties and liabilities are to be maintained.

One solution is to include such requirements in the tender specifications:

(i)   A contractor must include a method for remote commissioning utilizing suitably qualified regional staff, without limitation. This is typical in mainland countries. For example, in Australia, where one Author is supervising several other projects, qualified and experienced locally-based technical staff were able to enter into commercial arrangements with equipment OEMs during COVID-19 lockdowns to complete equipment commissioning inverters.

(ii)   While for most Pacific SIDS, the solution has been to delay commissioning during the pandemic, notably, after detailed discussion and planning, engagement with local technical staff, and using remote data and video, the TREP grid stability BESS on Tongatapu was recently commissioned remotely. This success demonstrates that remote commissioning is possible with proper consideration (though a high level of experience and qualification of local staff was also a precursor to this change). While there are cost implications, there are significant benefits.

(iii) A contractor must be able to service all equipment using personnel from a regional location (this requirement was introduced into later subproject specifications). Several major (BESS or inverter) equipment suppliers from Europe or the United States already have fully trained and qualified staff based in regional countries they service, including Australia, New Zealand, the United States, Fiji, and others, where the region is defined subject to the project country (for example, for Tonga or the Cook Islands, this may be South Pacific, New Zealand, or Australia—anywhere with direct flights). While this is likely to constrain competition for projects somewhat, it is not considered a showstopper and is necessary risk mitigation beyond COVID-19. Furthermore, as standardization of BESS and familiarity and/or experience of technicians continues to mature, it is expected that a move toward more generic installer requirements is likely.

Tonga, Tongatapu Solar PV IPP
(photo by Tonga Power Limited).

Key Recommendations for Future Projects

The recommendations from this assessment take into account the case study project challenges and successes, as identified through stakeholder consultations, analysis of project data, and comparison against industry practice. They focus on opportunities for improvement in project performance or for mitigating risk factors.

These recommendations are described below.

6.1 Project Design

(i) There are significant knowledge gaps for stakeholders about the associated technical issues, particularly with medium to high renewable energy hybrid systems, BESS, technology selection, and control systems requirements. Technical assistance consultants are not always able to understand or address the drivers and requirements of stakeholders. These factors contribute to reduced accuracy of risk assessments and suboptimal decision-making, which can be addressed through a consolidated program to build and maintain local energy literacy, supported by tools and information designed to clearly communicate key concepts. Some examples were presented in this report.

(ii)   Information to support decision-making is insufficient without a structured model for informed decision-making such as the responsible-accountable, consult, inform (RACI) model. For hybrid energy projects in isolated grids, stakeholder engagement during initial project selection and definition may benefit from using the following model:

    (a)   Sponsor: Allocate based on country structure (likely utility, government energy department)

    (b)   Government – energy ministry: Responsible and accountable (strategic) decision-maker

    (c)   Government – financial ministry: Consult

    (d)   Utility: Consult

    (e)   Regulator: Inform

    (f)   Customers: Consult and inform

    (g)   Landowners: Consult and inform

    (h)   Regional developers, contractors, and investors: Inform

    (i)   PMU, TA: Consult

(iii) Where possible, use of consistent delivery teams, including project management, administration, and technical assistance to support a unified decision-making team is also recommended.

6.2 Technology

(i)   Technology has matured substantially since the case study projects commenced and it is now apparent that BESS can offer a full suite of grid support functions allowing stable operation of small, medium, and large isolated networks with high renewable contribution, even without synchronous (diesel) generation online. However, there is still significant progress to be made, and particular gaps remain in

    (a)   product standardization;

    (b)   end-of-life treatment;

    (c)   clarity on emergency services response requirements; and

    (d)   consistency in definitions of control capabilities and in diesel-off operational capability.

(ii) For future projects, it is critical to understand these gaps. Noting that BESS products are not yet highly standardized, specifiers must give detailed consideration to the required project-specific functionality and operating environment and specify or select applicable standards or requirements accordingly.

(iii) It is also recommended to monitor ongoing technology advancement, including standards, and apply continuous improvement to technical specifications and concept development. However, presently, this may be demanding for SIDS and small utilities and should be supported through technical assistance or funding partners in the short to medium term.

6.3 Procurement

(i)   Procurement processes were identified as a challenge for many stakeholders. There was a desire to consider risk and opportunity through merit-based evaluation (particularly important for small, customized projects in remote locations, in a market with limited competition), or to manage complex, innovative projects. This approach is now facilitated in ADB's 2017 Procurement Policy that is applicable to all new projects using ADB financing. The following is recommended to allow better adaptation for hybrid projects:

  (a)   Undertake a comprehensive strategic procurement planning (SPP) exercise during the feasibility stage of a project in parallel with its technical development to identify an optimal procurement strategy that will deliver value-for-money outcomes.

  (b)   Include merit-point assessment criteria in the evaluation of all complex tenders, as standard, unless the SPP exercise determines it to not be the most suitable approach.

  (c)   Consider all available contracting modalities (e.g. Early Contractor Involvement – ECI) when developing the procurement strategy, ensuring that the modality chosen is best suited to the project and will facilitate effective competition.

It is recommended that the project delivery team engage early with ADB to utilize the flexibilities inhere in the 2017 Procurement Policy.

(i)   Additionally, given the nonstandard nature of projects to date and relatively high-risk exposure of the employer through to completion and commissioning, it is recommended to consider slight changes to performance securities and payment milestones. In particular, payment milestones should consider higher completion and commissioning payments. Performance securities should be maintained at a higher level through the first 2 years of operation, while battery degradation is verified. However, protections under the contract will also rely on strong project management processes that enable enforcement of the relevant protection measures.

(ii)   For projects requiring alignment, in particular where BESS or other utility-owned technology was deployed to support the connection of independent power producers (IPPs), it is considered most advantageous to plan for completing BESS at 3-9 months ahead of the IPPs. This was found to provide reasonable mitigation against the more significant risk of delaying IPPs commercial operations date. In many cases the BESS can still provide some project benefits prior to IPP connection.

(iii) Finally, in light of potential ongoing disruptions to travel for SIDS, it is recommended that all contracts contain a provision for remote commissioning and servicing from regional locations.

6.4 Insurance

(i)   Insurance options for Pacific projects are currently very limited. Considering their portfolio of investment and interest, an ADB or other funding agency backed insurance scheme may be a viable alternative. This would offer potential savings and reduce administrative time and cost in sourcing and negotiating insurance on a project by project basis. However, issues such as the impact on market competition for insurance services or fit with long-term operations period of projects would require careful consideration.

Tonga, Tongatapu, Power Station
Grid Stability BESS (Photo by TPL).

Summary of Interview Themes

Initial questions:

(ii)   Identify the project being addressed

(iii) What was your involvement in the project, during concept development, procurement and implementation?

(iv)   What do you believe was the objective of the project?

(v)   To what extent do you think this objective was achieved (not at all / somewhat / mostly / completely)

(vi)   Can you explain why you gave the above rating?

(vii) What do you think were the contributing factors to:

    (a)   Project success

    (b)   Issues that arose

(viii) Do you think there were other ways to achieve the project objective?

(ix)   Describe any key procurement or technical changes you think could improve project outcomes.

(x)   Any general thoughts.

Table A1: Summary of Interview Themes

	Themes Raised in Interviews	Multiple Projects	Multiple Respondents	Structural or Systemic Issue, Specific to this Project Type	Significant	Potential Future Opportunity or Risk	Relevant for Project Implementation Stage	Consolidate with Other Issues	Theme
Targets and performance	Renewable percentages not met due to demand changes	Y	Y	Y	Y	O	Y	Y	0
	Renewable percentages not met due to system design or performance issues	Y	Y	Y	Y	O	Y	Y	0
Stakeholders	Expectation that tariffs will go down	Y	Y	Y	Y	R	N	Y	1
	Lack of understanding of long-term O&M costs	Y	Y	Y	Y	R	N	Y	1
	Organizational adaptation to high renewables	Y	Y	Y	Y	O	Y	Y	1
	Supporting regulatory changes needed	Y	Y	Y	Y	R	Y	Y	1
	More customer focus needed and opportunities for participation such as rooftop systems	Y	Y	Y	Y	O	Y	Y	1
	Renewable energy fractions complicated by load growth	Y	Y	Y	Y	O	Y	Y	6
	More training and long-term support required	Y	Y	Y	Y	O	N	–	–
	Low justification for selection of some aspects of project selection and specification, decisions driven by grant funding access, political imperative or perception of decision makers	Y	Y	Y	Y	O	Y	Y	1

continued on next page

Table A1 *continued*

Themes Raised in Interviews	Multiple Projects	Multiple Respondents	Structural or Systemic Issue, Specific to this Project Type	Significant	Potential Future Opportunity or Risk	Relevant for Project Implementation Stage	Consolidate with Other Issues	Theme
Close contact and working relationship between contractors, technical advisors and utility added value	Y	Y	Y	Y	O	Y	Y	4
Procurement Inflexibility of ADB systems to account for local issues and qualitative factors in evaluation	Y	Y	Y	Y	O	Y	Y	2
Procurement errors	Y	N	N	N	R	Y	–	–
Generally strong support and protection from ADB following contractor complaints about award (when appropriate), however, not in one case	Y	Y	Y	Y	O	Y	Y	2
Lowest cost conforming tender is very difficult for these projects, particularly considering lack of product standardization and market evolution	Y	Y	Y	Y	O	Y	Y	2
Higher weightings for qualitative, quality, and support measures desired	Y	Y	Y	Y	O	Y	Y	2
More flexibility needed after the award stage for negotiation	Y	Y	Y	Y	O	Y	Y	2
Staging and coordination of different co-dependent contracts challenging	Y	Y	Y	Y	O	Y	Y	3

continued on next page

Table A1 *continued*

	Themes Raised in Interviews	Multiple Projects	Multiple Respondents	Structural or Systemic Issue, Specific to this Project Type	Significant	Potential Future Opportunity or Risk	Relevant for Project Implementation Stage	Consolidate with Other Issues	Theme
Contract management	Incentives for Contractors to perform and deliver on time inadequate or difficult to enforce	Y	Y	Y	Y	O	Y	Y	4
	Remoteness of Contractors and ability to supervise local works	Y	Y	Y	Y	O	Y	Y	4
	Generally strong support from ADB for changes in scope or financing as project develops	Y	Y	Y	Y	O	Y	Y	4
	More reactive processing of invoices or project change requirements	Y	Y	Y	Y	O	Y	Y	4
Technology	Basis for preselection of best technology for the task unclear	Y	Y	Y	Y	O	Y	Y	6
	Technology change and new capabilities and/or lower cost	Y	Y	Y	Y	O	Y	Y	5
	Technology change obselesence	N	N	Y	Y	R	N	–	–
	Understanding requirement for stability vs load-shifting	Y	Y	Y	Y	O	Y	Y	6
	Understanding of integration requirement	Y	Y	Y	Y	O	Y	Y	3, 6
	Reliance on consultants	Y	Y	Y	Y	O, R	Y	Y	6?
	Technology introducing new challenges in traditional network (harmonics, impedence, fault levels, protection, etc.)	Y	Y	Y	Y	O	Y	Y	3, 6

continued on next page

Table A1 continued

Themes Raised in Interviews	Multiple Projects	Multiple Respondents	Structural or Systemic Issue, Specific to this Project Type	Significant	Potential Future Opportunity or Risk	Relevant for Project Implementation Stage	Consolidate with Other Issues	Theme
Uncertainty if things will work	Y	Y	Y	Y	O	Y	Y	6
Complexity to manage distributed systems	Y	Y	Y	Y	O	Y	Y	6
Miscellaneous issues with building leaking, corrosion, electrical interference, inadvertent trips	Y	Y	N	Y	O	Y	–	–
Bespoke systems or implementations problematic	Y	Y	Y	Y	O	Y	Y	2
Modular, mature systems desired	Y	Y	Y	Y	O	Y	Y	2
Long delays for spares	Y	Y	Y	Y	O	N	–	–
Consistencey in technical support added value	Y	Y	N	Y	O	Y	–	–
Comprehensive tender specification added value	Y	Y	Y	Y	O	Y	Y	2
Improve basis for sizing BESS	Y	Y	Y	Y	O	Y	Y	6
Include control system requirements for diesel generators	Y	Y	Y	Y	O	Y	Y	3, 6
COVID-19								
Insurance	Y	Y	Y	Y	R	Y	N	7
Project Delay	Y	Y	Y	y	R	Y	Y	7
Remote Commissioning	Y	Y	Y	Y	R	Y	N	8
Other								
Alternative financing models such as IPP could be considered	Y	Y	Y	Y	O	N	–	–

BESS = battery energy storage system, IPP = independent power producer.

Source: Author.

Cook Islands, Mangaia Hybrid Power Station BESS (Photo by Entura).

APPENDIX 2
Basic Performance of Subprojects

While not a focus of this analysis, consistent with the analysis methodology parts 1 and 2 (see section 4.2), subproject delivery was assessed using a range of measures related to project implementation, from concept to initial operation. The results are shown in Table A2.

Table A2: Subproject Performance Assessment (delivery to plan and technical performance)

	OIREP		TREP				CIRESP				
	4 micro-grids	3 mini-grids	PS BESS	LS BESS	2 x mini-grid	5 x micro-grid	GEF BESS	GCF BESS GS	GCF BESS LS	1 x mini-grid	4 x micro-grids
Subproject concept changed significantly prior to feasibility	N	N	Y1	Y1	N	N	N	N	N	N	N
Subproject concept changed prior to procurement	N	N	N	N	N	N	N	Y3	N	N	N

continued on next page

Table A2 *continued*

	OIREP		TREP				CIRESP				
	4 micro-grids	3 mini-grids	PS BESS	LS BESS	2 x mini-grid	5 x micro-grid	GEF BESS	GCF BESS GS	GCF BESS LS	1 x mini-grid	4 x micro-grids
Subprojects procured within budget (including contingency)	N	Y	Y	Y	Y	Y	Y	Y	Y	Y	Y
Subprojects completed to specification	N	Y	TBC	TBC	TBC	TBC	Y	Y	Y	Y	Y
Minor or moderate defects rectified during DLP	Y	Y	TBC	TBC	TBC	TBC	Y	Y	Y	Y	Y
Major defects, or defects not resolved in DLP	TBC	Y4	TBC	TBC	TBC	TBC	Y5	N	N	N	N
Performance guarantees substantially met	TBC	Y	TBC	TBC	TBC	TBC	Y	TBC	Y	Y	Y
Network impact as per feasibility	TBC	Y10	TBC	TBC	TBC	TBC	Y	TBC	TBC	Y	Y
RE benefit as per feasibility	TBC	N11	TBC	TBC	TBC	TBC	N6	TBC	TBC	Y	N7
Period from concept to Contract effectiveness within 6 months of initial schedule	N8	N8	N8	N8	N8	N8	N8	N8	N8	N8	N8
Period from Contract Effectiveness to completion within 3 months of contract schedule	N8	N8	N9	N9	N9	N9	N9	N9	Y	Y	N9

BESS = battery energy storage system, CIRESP = Cook Islands Renewable Energy Sector Project, DLP = defects liability period, GCF = Green Climate Fund, GEF = Global Environment Facility, GS = grid stability, LS = load-shifting, TBC = to be confirmed, TREP = Tonga Renewable Energy Project.

Notes:
1. Changed from distributed to centralized BESS.
2. Not used.
3. Changed from load-shifting the grid stability BESS (power increased and storage decreased).
4. Various faults attributable to bespoke implementation on one project.
5. A number of minor to moderate, plus replacement of battery cells initiated by supplier (not defect) were outstanding at start of coronavirus disease (COVID-19) travel restrictions and have not yet been rectified due to access.
6. Pending installation of control system by utility and additional solar photovoltaic.
7. RE 80%–92% compared with 90%–95% at feasibility. Attributed to demand growth, and some operational issues (rectified) during DLP.
8. Delays attributable to various issues in the context of aggressive schedules, including reprocurement, extended approvals periods from multiple parties, minor errors, and administrative issues.
9. Delays typically due to either one, reasonable extension of time granted due to COVID-19 delays or interface between multiple contracts, or two, lack of appropriate contractor resourcing allocation to the project.
10. Performed as expected except for the various faults noted in the defects.
11. Lower than expected renewable energy due to load growth.
Source: Author.

In general, projects experience only small to moderate change between conception and procurement and were substantially completed to the procurement specification. Most projects had minor to moderate defects that were typically rectified during the defects liability period, though some extended longer than this period due to COVID-19 related travel restrictions—the majority of projects performed in accordance with their functional guarantees, some with minor discrepancies.

The most significant discrepancy with respect to expectation was delivery to schedule, with all projects taking longer to reach contract award than planned initially and all projects experiencing some delay during contract execution.

Delays prior to contract award had various causes, including additional time taken in procurement during evaluation or approval of award than expected; administrative delays in achieving contract effectiveness (such as arranging advance payments, securities, letters of credit); complex decisions about the project procurement structure, and finalizing the land acquisition. Delays during the execution were commonly attributable to COVID-19, the interface between contracts, or in some cases, inadequate resourcing of Contractors.

Cook Islands, Rarotonga, Hospital Rooftop
Solar System (Photo by Te Aponga Uira [TAU]).

References

Asian Development Bank. 2021. *Cook Islands: Renewable Energy Sector Project.* https://www.adb.org/
    projects/46453-002/main (accessed 5 October 2021).

Australian Broadcasting Corporation. 2021. *Fire at Tesla Giant Battery Project Near Geelong was Likely Caused
    by Coolant Leak, Investigation Finds.* https://www.abc.net.au/news/2021-09-28/fire-at-tesla-giant-
    battery- project-near-geelong-investigation/100496688 (accessed 10 November 2021).

T. Briscoe, C. Baccineli, and J. Chambless. 2000. *Best Practices—Project Review Process.* Houston: Newtown Square.

G. Chaudhary et al. 2021. Review of Energy Storage and Energy Management System Control Strategies
    in Microgrids. *Energies.* 14 (4929).

S. Cherevatskiy. et al. 2020. *Grid Forming Energy Storage System Addresses Challenges of Grids with High Penetration
    of Renewables (A Case Study).* Paris: CIGRE.

T. Costello. 2012. RACI—Getting Projects "Unstuck". *IT Professional.* 14 (2). pp. 64–66.

DNV-GL. 2019. *Development of a Proposed Performance Standard for a Battery Storage System Connected to a Domestic / Small Commercial Solar PV System: Gap Analysis of Existing Battery Energy Storage System Standards.* ISSU. https://issuu.com/dnvgl/docs/190624090258-6c860998b91c42d29b20800cd61707cf/1.

Entura. 2015. *Cook Islands Renewable Energy Project Atiu Subproject Feasibility.* Cook Islands Ministry of Finance and Economic Management. http://www.mfem.gov.ck/development/development-partners/asian-development-bank#project-reports Asian Development Bank - Cook Islands - Ministry of Finance and Economic Management (mfem.gov.ck).

S. George and C. Egbu. 2016. Modern Selection Criteria for Procurement Methods in Construction. *International Journal of Managing Projects in Business.* 9 (2). pp. pp 309–337.

Government of Tonga. 2010. *Tonga Energy Road Map (2010–2020).* Nukalofa: ARUP.

G. Hering. 2019. *Burning Concern: Energy Storage Industry Battles Battery Fires.* https://www.spglobal.com/marketintelligence/en/news-insights/latest-news-headlines/ burning-concern-energy-storage-industry- battles-battery-fires-51900636 (accessed 1 November 2021).

S. A. Hirmer et al. 2021. Stakeholder Decision-Making: Understanding Sierra Leone's Energy Sector. *Renewable and Sustainable Energy Reviews.* Volume 145.

International Electrotechnical Commission. 2019. *Secondary Cells and Batteries—Marking Symbols for Identification of Their Chemistry.* Geneva.

D. Leitch. 2020. *Virtual Inertia in Practice: How South Australia's Second Big Battery Made Its Mark.* https://reneweconomy.com.au/virtual-inertia-in-practice-how-south-australias-second-big-battery-made-its-mark-12934/ (accessed 20 December 2021).

Lin, Y. et al. 2020. *Grid Forming Inverters. Golden, CO.* National Renewable Energy Laboratory. https://www.nrel.gov/docs/fy21osti/73476.pdf.

M. B. McKinnon, S. D. S. K., 2020. *Four Firefighters Injured In Lithium-Ion Battery Energy Storage System Explosion—Arizona.* Columbia: Underwriters Laboratories Inc.

Ministry of Finance and Economic Management, 2017. *Cook Islands Renewable Energy Charts and Implementation Plans.* http://www.mfem.gov.ck/447-cook-islands-renewable-energy-chart-planning (accessed 3 November 2021).

———. 2021. *Southern Renewable Energy Project.* http://www.mfem.gov.ck/development/development-partners/asian-development-bank (accessed 7 October 2021).

A. Muriro and G. Wood. 2010. *A Comparative Analysis of Procurement Methods Used on Competitively Tendered Office Projects in the UK.* University of Salford, Manchester. http://usir.salford.ac.uk/id/eprint/23055/

D. Nikolic et al. 2016. Cook Islands: 100% Renewable Energy in Different Guises. *Energy Procedia.* Volume 103. pp. 207–212.

Organisation for Economic Co-operation and Development. 2015. *Small Island Developing States (SIDS) and the Post-2015 Development Finance Agenda.* https://www.oecd.org/dac/financing-sustainable-development/Addis%20Flyer%20SIDS%20FINAL.pdf.

D. Pattabiraman, R. H. Lasseter, and T. M. Jahns. 2018. Comparison of Grid Following and Grid Forming Control for a High Inverter Penetration Power System. *IEEE.* pp. 1–5. doi: 10.1109/PESGM.2018.858616.

E. Samoglou. 2014. *Fuel Shortage in Cook Islands May Become Crisis.* http://www.pireport.org/articles/2014/05/29/fuel-shortage-cook-islands-may-become-crisis (accessed 5 May 2022).

State Government of Victoria. 2021. *Innovation and the Procurement Process—Goods and Services Procurement Guide.* https://www.buyingfor.vic.gov.au/innovation-and-procurement-process-goods-and-services-procurement-guide (accessed 12 November 2021).

F. Syme-Buchanan. 2019. *Call for Cook Islands to Diversify Its Economy.* https://www.rnz.co.nz/international/pacific-news/380530/call-for-cook-islands-to-diversify-its-economy (accessed 2 November 2021).

Te Aponga Uira. 2019. *Rarotonga Renewable Energy Program Summary.* Paper presented in the 28th Pacific Power Association Conference. Rarotonga.

United Nations. 2021. *List of SIDS.* https://www.un.org/ohrlls/content/list-sids (accessed 2 November 2021).

United Nations Development Programme. 2021. *Towards a Multidimensional Vulnerability Index.* https://www.undp.org/publications/towards-multidimensional-vulnerability-index (accessed 3 November 2021).

Utilities Regulatory Authority. 2019. *Comparative Report: Pacific Region Electricity Bills,* Vanuatu.

M. A. E. Wardani, J. I. Messner, and M. J. Horman. 2006. Comparing Procurement Methods for Design–Build Projects. *Journal of Construction Engineering and Management.* 132 (3).

Lightning Source UK Ltd.
Milton Keynes UK
UKHW050914160922
408923UK00005B/268